Robotics in Horticulture

NIPA® GENX ELECTRONIC RESOURCES & SOLUTIONS P. LTD.
New Delhi-110 034

About the Editors

Ms. Ikra Manzoor has completed her B.Sc. Horticulture (Hons.) degree from Sher-e-Kashmir University of Agricultural Sciences and Technology of Kashmir in 2018. She did her M.Sc. Horticulture (Fruit Science) from Sher-e-Kashmir University of Agricultural Sciences and Technology of Kashmir in 2021. She was a merit scholarship holder during both the degree programs. Currently she is pursuing her Ph.D. in Fruit Science from Sher-e-Kashmir University of Agricultural Sciences and Technology of Kashmir. She was shortlisted for zonal level evaluation for Agri-Hackethon of ICAR in 2020 for the technological innovative idea in the field of horticulture. She has qualified ASRB-NET (National Eligibility Test) & ASRB-ARS (Mains) in the discipline of Fruit Science in UR category in 2021. She has also published abstracts on diverse topics related to horticulture in various National & International seminars and symposiums.She has also contributed as an author for several book chapters published in Taylor & Francis. She has also won the best poster presentation award at the 9th Indian Horticulture Congress in 2021. She was selected for the award of Prime Minister's Doctorate Fellowship in 2022. She has authored a book in the field of Horticulture in 2022. Presently, she published various papers in highly reputed journals like *Frontiers in Plant Science* with NAAS rating (12.63) & IF (6.627), Journal of Integrative Agriculture IF (4.8) and Journal of Environmental Chemical Engineering IF (7.3) from the 2023 to 2024. She has also received Young Researcher Award in the "International Conference on New Generation Horticulture for Prosperity" organized by Orrisa Horticultural Society, College of Agriculture, OUAT, Bhubaneswar-Odhisa, India in 2023. She has published a book in the field of Tissue Culture in the year 2024.

Dr. Shabir Hussain Wani Assistant Professor (Senior Scale) cum Scientist, Genetics and Plant Breeding at Mountain Research Centre for Field Crops, Khudwani, Sher-e-Kashmir University of Agricultural Sciences and Technology, Srinagar J&K, India. He has published more than 200 papers/chapters in journals and books ofinternational and national repute. He served as guest editor and reviews editor for the Journal Frontier in Plant Science (2015-2018). Presently he is also serving as Associate Editor for PCTOC, Springer and reviews editor for Frontiers in Plant Science, section Abiotic Stress in Plants. He has also edited several books on current topics in crop improvement for abiotic stress tolerance published bySpringer Nature and CRC press USA. He was elected as Fellow of Linnean Society of London 2017. Recently he was felicitated with Award for Excellence in Research 2020 from Novel Research Academy, Puducherry, India. Recently Dr. Wani was selected as member of Expert Working Group (EWG) Wheat Initiative for Adaptation of Wheat to Abiotic Stress (AWAS) and Wheat Breeding Methods and Strategies (WBMS), Berlin, Germany during January 2021 and 2022 respectively. He has been judged as Fellow of International Society for development and Sustainability ISDS, Japan. Recently, he was also selected as member. The Royal Society of Biology, London, UK. Dr Wani was selected as Member, National Agricultural Sciences India (NASI), 2022.

Robotics in Horticulture

Editors

Ikra Manzoor
Shabir Hussain Wani

NIPA® GENX ELECTRONIC RESOURCES & SOLUTIONS P. LTD.
New Delhi-110 034

NIPA® GENX ELECTRONIC RESOURCES & SOLUTIONS P. LTD.

101,103, Vikas Surya Plaza, CU Block
L.S.C. Market, Pitam Pura, New Delhi-110 034
Ph : +91-11-43860225, Mob.: +91 9717133558, 9540816132
E-mail: newindiapublishingagency@gmail.com
Website: www.nipaersources.com

Print ISBN: 978-93-58875-05-8
ebook ISBN: 978-93-58877-79-3

Composed and Designed by NIPA®.

This book is dedicated
to my beloved papa

Mr. Manzoor Ahmad Bhat

who is my mentor and spiritual role model
who breath the last on 11 July 2024

Preface

Artificial intelligence AI is a general purpose technology with significant implications for all aspects of society and everyday life. In different sectors, such as health care, manufacturing, automobile, finance and agriculture, AI technology is being applied in order to increase productivity and efficiency and overcome the usual challenges. Understanding intelligence so that it can be reproduced in machines is a prerequisite for the development of artificial intelligence. Most AI systems collect input, analyze this data and make decisions on the basis of these inputs. Embracing AI yields numerous positive outcomes and societal advantages. Its applications hold the potential to enhance living standards, promote health, facilitate judicial processes, foster economic prosperity, bolster public safety, and mitigate human impact on the environment and climate. AI acts as a valuable ally, empowering individuals to complete tasks with greater efficiency and effectiveness, leading to a multitude of advantages. Furthermore, AI goes beyond simply optimizing current tasks; it opens doors to new possibilities, such as analyzing vast research data sets, paving the way for groundbreaking scientific discoveries that could positively impact every aspect of our lives. In today's agriculture, drones and robots are like our high-tech helpers, wandering the fields and gathering data to make farming smarter. With the help of interconnected technologies and intelligent analytics, farmers can now comprehend what's occurring in real-time on their farms. They're learning how to grow more crops, use resources more efficiently, manage their farms better, and make eco-friendly choices about where and how to use those resources.This book has been prepared to provide comprehensive knowledge about role of robotics in horticulture including various topics viz., Artificial Intelligence, Crop and Soil Monitoring in Horticulture Crops, Robotics in Training and Pruning of Horticultural Crops, Role of AI in Diagnosis and Monitoring of Diseases and Insect Pests, Intelligent spraying in horticultural crops, Robotic Pollination in Horticultural Crops, Future Of AI In Horticultural Crops. All these techniques and their various applications have been thoroughly discussed in this book. This book provides a thorough understanding of these chapters and their role in the development of horticulture crops this book has been divided into seven thoroughly dealt chapters that provide all the novel information required.

We solemnly believe that it will provide all the necessary and important information to students, research scholars, research institutes and establishments involved with R&D of agricultural crops.

Content

1

Artificial Intelligence

Nowsheen Noor and Syed Zainab Kashani

Division of Vegetable Science, Science Sher-e-Kashmir University of Agricultural Sciences and Technology of Kashmir, Shalimar, Srinagar-190025, Jammu and Kashmir, India

Abstract

Artificial intelligence AI is a general purpose technology with significant implications for all aspects of society and everyday life. In different sectors, such as health care, manufacturing, automobile, finance and agriculture, AI technology is being applied in order to increase productivity and efficiency and overcome the usual challenges. Understanding intelligence so that it can be reproduced in machines is a prerequisite for the development of artificial intelligence. Most AI systems collect input, analyze this data and make decisions on the basis of these inputs. Embracing AI yields numerous positive outcomes and societal advantages. Its applications hold the potential to enhance living standards, promote health, facilitate judicial processes, foster economic prosperity, bolster public safety, and mitigate human impact on the environment and climate. AI acts as a valuable ally, empowering individuals to complete tasks with greater efficiency and effectiveness, leading to a multitude of advantages. Furthermore, AI goes beyond simply optimizing current tasks; it opens doors to new possibilities, such as analyzing vast research data sets, paving the way for groundbreaking scientific discoveries that could positively impact every aspect of our lives. In today's agriculture, drones and robots are like our high-tech helpers, wandering the fields and gathering data to make farming smarter. With the help of interconnected technologies and intelligent analytics, farmers can now comprehend what's occurring in real-time on their farms. They're learning how to grow more crops, use resources more efficiently, manage their farms better, and make eco-friendly choices about where and how to use those resources.

Keywords: *Artificial intelligence, optimization, research, drones, robots, analytics, eco-friendly, biotechnology.*

1.1 Introduction

Since the beginning of humankind, we have thrived as a species due to our unique capabilities. These skills have developed and progressed over the centuries to the point that, in recent decades, we have been able to build machines that can imitate our intelligence, as far as we comprehend it. A novel type of world may lie ahead of us.

Artificial intelligence is the future, and the future is here." - Dave Waters

Artificial intelligence (AI) refers to the replication of human intelligence in computers, which are programmed to think and behave in a manner similar to humans. Literacy, logic, problem-solving, perception, and linguistic appreciation are all manifestations of cognitive abilities. Artificial Intelligence refers to the development of computer systems, computer-controlled robots, or software that are designed to exhibit intelligent behavior similar to that of the human mind. The term artificial intelligence is composed of the word "Artificial," which denotes something created by humans rather than existing naturally, and "Intelligence," which refers to the capacity to gain and utilize knowledge and skills.AI is achieved by the analysis of patterns in the human brain and the evaluation of cognitive processes. The culmination of these studies leads to the creation of sophisticated software and systems with advanced intelligence. The term AI was coined by McCarthy *et al.* 2006 while working on Dartmouth summer research project in which they proposed that machines have the capability to emulate "every facet of learning or any other characteristic of intelligence."

If a machine could demonstrate human-like traits like flexibility in problem-solving and decision-making, the capacity to judge the relative importance of various items, the ability to tell apart objects that appear similar, and vice versa, then the machine is regarded as intelligent. Contrary to natural intelligence, a machine can only be endowed with intelligence within a specific and limited field. Inherent intelligence is more universal and exhibits a significant degree of adaptability. Clearly, it is unattainable to have computers with such extraordinary intelligence. AI encompasses the field dedicated to creating intelligent machines and software. The primary objective of AI is to design computers to execute specific activities in a manner that mimics human capabilities, including information acquisition, logical reasoning, learning, planning, and problem-solving. Machine Learning and Deep Learning are fundamental components of AI (Patel *et al.*, 2020a, 2020b; Pandya *et al.*, 2019; Sukhadia *et al.*, 2020). While ML refers to the capacity of systems to learn without explicit programming. The primary attribute of machine learning is the ability to acquire knowledge through experience in order to address a wide range of problems. DL specifically involves the training of deep neural networks, according to Kodali and Sahu (2016) and Kulkarni and Deshmukh (2013). Deep learning models are composed of numerous layers that can turn low-level representations into higher-level, more abstract representations using non-linear transformations (LeCun *et al.* 2015).Currently, there are various deep learning algorithms that are widely used in different fields such as engineering, bioinformatics, agriculture, and medical science(Fusco *et al.* 2021).

1.2 Beginning of AI

The beginnings of Artificial Intelligence (AI) are multifaceted, with different views on when it actually started. Some think it began with philosophical questions about human nature, while others point to math advancements or the creation of computers. Alan Turing is often recognized as the progenitor of artificial intelligence and contemporary computer science due to his early recognition and formulation of principles regarding intelligent machines. He was an early advocate of the idea that the operations of the human brain are fundamentally analogous to those of a digital computing system. He played a key role in the pioneering experiment famously known as "The Turing Test," which marked a significant moment in the advancement of AI. His influential work, 'Computing Machinery and Intelligence,' explored the intriguing idea of whether a computer devoid of life could demonstrate human-like cognitive processes. This publication represented a noteworthy achievement in the domain. Basically, if a machine fools a person into thinking it's human, it passes the test. Even though Turing's work was groundbreaking, AI's development has gone through various stages and changes. While the Turing test sparks interest, some argue it's not the best measure of AI (Halpern, 2006). Critics say it doesn't truly assess AI's ability to think or problem-solve independently, as it mainly focuses on a machine's skill at fooling humans. They also note that important aspects like creativity and empathy are left out (Stahl, 2021b). However, supporters of the Turing test see it as a useful tool for comparing machine and human intelligence, pinpointing where machines excel and where they fall short (Liao, 2020). Yet, it's important to remember that the Turing test isn't the only way to gauge AI. Other methods, like task-based or behavior-based assessments, offer different perspectives (Bryson, 2020). In the end, evaluating AI requires looking at its diverse capabilities and considering how they align with our understanding of intelligence. It's about taking a broader, more inclusive approach.

1.2.1 Fundamental Features of AI

The goal of artificial intelligence (AI) is to build computer systems that are capable of handling complex tasks that usually need human-like skills. These problems frequently entail high levels of complexity or natural tasks that are challenging for traditional algorithmic approaches to handle, including image recognition or natural language comprehension. As is frequently noted in computer science, AI algorithms primarily handle symbolic data instead of just numerical data.

Knowledge is essential in the field of artificial intelligence (AI), and it is emphasized that information must be acquired, represented, and used in a particular domain. Knowledge-based systems are a major area of interest for AI research and development, highlighting the need of successfully integrating domain-specific knowledge.AI is inherently interdisciplinary, drawing insights from a variety of fields including computer science, logic, linguistics, and cognitive sciences. Furthermore, a variety of industries, including healthcare, manufacturing, banking, and finance, use AI.

1.2.2 Components of AI

Learning: Learning in AI mirrors human knowledge acquisition but is implemented computationally. One key aspect is trial and error, where the AI system tries different approaches to a problem and retains successful strategies for future use.Another learning method is rote learning, where the AI memorizes specific information, like problem-solving techniques or vocabulary, and applies it across various contexts. For instance, in image recognition, AI learns to identify objects by analyzing patterns in a dataset.

Reasoning and Decision Making: AI analyzes information and makes decisions through reasoning, drawing inferences from situations. This can be deductive, providing guaranteed conclusions, or inductive, dealing with uncertain outcomes. For example, chess programs use reasoning to evaluate possible moves and predict outcomes.

Problem Solving: AI's problem-solving ability involves techniques like planning, search, and optimization, tailored to specific problems or applicable to diverse issues. For instance, route optimization algorithms in navigation systems find the most efficient path between two points.

Perception:AI perceives its environment through sensors, interpreting data to understand scenes, objects, and relationships. This is vital for tasks like computer vision and speech recognition.Self-driving cars, for example, use perception to navigate roads by recognizing obstacles and traffic signals.

Language Processing: Language processing in AI involves understanding and generating natural language, enabling effective interaction. Techniques like natural language understanding and machine translation support applications like chatbots and sentiment analysis. Virtual assistants, for instance, comprehend and respond to spoken commands, showcasing AI's language processing capabilities.

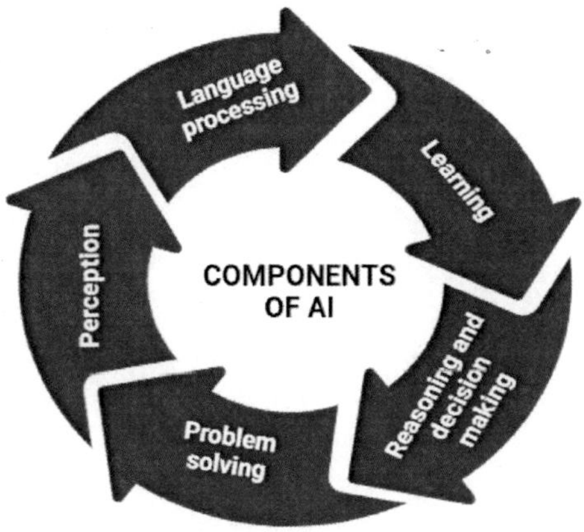

Fig. 1.1 Components of AI

1.2.3 Sub Fields of AI

Artificial Intelligence (AI) is a broad field that involves creating intelligence in computers or making them act smart like humans. Within AI, there are many subfields, each focusing on different aspects of intelligence. The main ones are:

1. Machine Learning: This involves teaching computers to learn from data and make predictions without being explicitly programmed. It's like teaching them to learn from experience.
2. Natural Language Processing (NLP): This helps computers understand and interact with human language, like speech recognition or virtual assistants such as Siri and Alexa.
3. Expert Systems: These are like specialized databases that contain expert knowledge in a specific field. They're used to help people make decisions or solve problems in that area.
4. Computer Vision: This is about teaching computers to "see" and understand images or videos. It's used in things like facial recognition or quality control in manufacturing.

5. Robotics: This involves creating and programming robots to perform tasks, like walking or driving, using technologies from other AI subfields like computer vision and machine learning.

These subfields often overlap and work together to create intelligent systems. Each plays a role in different applications, from everyday devices like smartphones to complex machines like self-driving cars.

1.2.4 Classification of AI

According to the degree of cognitive ability, artificial intelligence (AI) can be categorized into various types.

1. Weak or limited artificial intelligence: Also referred to as "functional" AI, this type of AI is specifically engineered to execute particular tasks efficiently. However, it lacks the capability to reason in a generalized manner or learn from novel situations.Weak or limited AI denotes a form of AI restricted to performing specific tasks without the capacity to adjust to unfamiliar circumstances. It is tailored for solving precise problems and lacks the ability to generalize its actions across different scenarios (Chen & Chen, 2022). Weak AI finds application in systems requiring repetitive tasks, such as email spam filtering, information categorization, content recommendations on streaming platforms, and product suggestions on e-commerce platforms.
2. General or strong artificial intelligence (AGI) possesses the capability to undertake a wide array of cognitive tasks, including reasoning, learning, and problem-solving, while also being adaptable to novel situations and environments. This contrasts with weak or limited AI, which is confined to specific tasks and lacks adaptability. The term "general AI" stems from its focus on creating computing systems capable of performing diverse tasks, aiming to emulate human intelligence as closely as possible (Stahl, 2021a).An illustration of general AI lies in advanced personal assistant systems capable of answering questions and executing tasks based on user preferences, exemplified by the GPT3 chatbot's ability to generate text in a highly logical and adaptive manner (Katrak, 2022). This enables general AI to adapt to varying scenarios and provide personalized assistance.
3. Artificial superintelligence (ASI), also termed "high-performance" AI, possesses the capacity to execute virtually any task demanding human intelligence and can surpass humans in cognitive and learning abilities (Tzimas, 2021). This immensely powerful form of AI excels in performing intricate tasks with exceptional precision and efficiency, commonly employed in domains requiring extensive data processing,

such as data science and medical research. An instance of high-performance AI includes a computer system suitable for swiftly and accurately processing and analyzing vast volumes of data. For instance, the navigation system of Tesla vehicles capable of autonomous driving exemplifies the capabilities of high-performance AI. Additionally, high-performance AI is utilized in data science systems to analyze extensive datasets and extract valuable insights for decision-making across diverse fields, such as collective facial recognition systems. In medical research, high-performance AI plays a crucial role in analyzing large medical datasets to aid scientists in discovering novel treatments for diseases.

1.3 AI in Agriculture Including Horticulture

The benefits of artificial intelligence in different sectors of the global economy are brought by its rapid development, fueled by advances in computational power and increased use of cloud computing. In all sectors, artificial intelligence technologies help people overcome the conventional obstacles. Some sectors that use AI applications are finance, transport, health care and agriculture (Shadrin *et al.* 2019).The ongoing reliance on traditional farming methods has kept Indian agriculture operating at a subsistence level. Recent shifts in climate patterns present additional risks to food security and nutritional well-being (Kumar and Joshiba2019). Given the growing population and impending challenges, the adoption of automation in horticulture appears to be essential (Chaudhry and Garg 2019). The use of AI and ML technologies in particular is having a significant impact on agriculture. Smart farming practices are being developed with AI to reduce farmers losses while providing them with high yields (Liu 2020; Talaviya *et al.* 2020; Ben Ayed and Hanana 2021). Businesses utilize satellite imagery and meteorological data to ascertain land area and monitor the health of crops in real-time (Subeesh and Mehta 2021; Vijayakumar and Balakrishnan 2021). The widespread adoption of AI in agriculture holds significant promise and potential for innovation (Mohr and Kühl 2021; Sharma 2021). The AI powered sensors are identifying weed infested areas and prescribing the correct treatment of herbicides. In addition, artificial intelligence systems are predicting weather patterns, assessing crop health and detecting diseases, pests and deficiencies in nutrition. Crop health monitoring and pest control assistance is provided by drones with artificial intelligence capabilities. The sensors can give information on the various ripening phases of fruits and vegetables. To reduce the workload of workers, robots are increasingly used to perform tasks that require a high degree of automation and reduction in labour costs. Automation is becoming essential to meet production demands as the world's population increases and demand for farming products rises. AI is facilitating the use of predictive analytics, improved farm management and increased crop quality assurance. In order to enhance their accuracy and performance, AI and IoT sensors can be added to the machines. These methods help to assess individual plots or plants with greater precision in terms of both space and time. AI may simplify the process of finding suitable targets for genetic modification in large genome data and creating efficient artificial promoters to enhance agronomic qualities in plants (Pandey andChaudhary 2016, 2021).Combining comprehensive image-based features with omics data in agriculture may help discover critical traits associated with stress tolerance and acclimatization mechanisms (Marchetti *et al.*, 2019), hence facilitating the development of crops that are resistant to climate change. With AI, farmers gain insights into market demand and consumer preferences, enhancing their returns. AI-driven supply chains streamline logistics and reduce costs, benefiting smaller farmers by getting their products to market faster and with less waste. The number of challenges faced by growers can be addressed through automation. Artificial intelligence helps farmers to select the right crop, depending on climatic and soil conditions, which can lead to high yields. In the identification of pests and pathogens, AI gives 98% accuracy.

1.3.1 Applications in Horticulture

- **Prediction of Weather**

The horticulture business relies heavily on weather forecasting because many inter-cultural operations depend on it. Using algorithms in conjunction with satellites, machine learning technology can be useful in making daily weather predictions. These algorithms have the ability to additionally be utilized to forecast crop health (Anonymous 2019) without coming into contact with big fields. The range of pest and pathogen prediction and detection in a farmer's field is further expanded by machine learning.

- **Soil Health Assessment**

In order to determine the moisture and nutrient content in the soil, algorithms and sensors developed using both artificial intelligence and machine learning may be used. Besides this, robots can also be employed for the measurement of nutrients in the soil that could also offer predictive analysis for the ideal amount of irrigation and fertilizer. Soil health monitoring without the requirement of laboratory testing is now possible due to image recognition and deep learning models. Artificial intelligence (AI) systems that are combined with data signals from distant satellites and local picture capture on farms enable farmers to take the quickest feasible action to improve the health of their soil.

• Crop Health Monitoring

To sustain agricultural productivity, it is essential to evaluate crop health and identify infestations as soon as possible. For effective mitigation, stressors like as pest and disease infestations, nutrient deficits, and climate change must be identified in a timely manner. The utilization of visual analysis for the detection of diseases and pests has become the prevailing method for automatically diagnosing plant stress. This automatic detection method leverages advanced deep learning-based AI techniques, decreasing the requirement for human interaction. Numerous efforts have been made to diagnose both illnesses and insect pests in crops using deep learning approaches. In order to identify and categorize the photos of diseases and pests of tomato crop, Liu and Wang (2020) employed theYOLO-v3 object detector model. Cheeti *et al.* (2021) introduced a method for detecting insect pests in crops by employing sophisticated deep learning methodologies. Their approach involved utilizing the YOLO (You Only Look Once) algorithm for identifying and pinpointing the pests within images, alongside employing the AlexNet model as a classifier for image categorization.

• Field Management

There are several difficulties with manually applying pesticides, such as using too many chemicals, not having enough workers, distributing the spray unevenly, contaminating the environment, and having restricted coverage. Drones and other advances in aerial systems, however, present a possible remedy. Drones' high-definition photos offer instantaneous evaluations of regions in need of field-applied pesticides.

Hafeez *et al.* (2022) provided a comprehensive overview of the application of unmanned aerial vehicles (UAVs) for the purpose of spraying pesticides in agricultural areas. Their study emphasizes that the use of drone-mounted sprayers has led to notable enhancements in coverage, increased efficiency of chemical application, and optimized the overall spraying procedure, resulting in improved convenience and speed.

• Plant Phenotyping

Plant phenotyping is the term used to describe a group of methods and procedures used to precisely and accurately assess different characteristics of the growth, structure, and composition of plants at different organizational levels, from individual organs to entire crop canopies. Applications in plant production are included in the term, even though it is often linked to plant breeding goals. One prevalent approach in digital phenotyping consists of utilizing image analysis, where cameras are used to capture images and software is employed to automatically derive measurements from these images. This technique enables the examination of plant morphology consistently and accurately.

- **Smart Irrigation Systems**

Smart irrigation aims to supply the appropriate amount of water to the appropriate crop at the appropriate time and location to increase the yield of crops. Issues related to India's low irrigation efficiency (around 38%) can be addressed more fully by using sensor-based automated irrigation systems. Additionally, machines trained on past weather patterns, the local soil condition, and the type of crops to be cultivated can be used to schedule irrigation automatically.

- **Assisting Pollination**

The decline in natural pollinators, which has an impact on food production, is being addressed by exploring artificial intelligence. Researchers at Monash University used AI and Computer Vision to track the movements of various pollinators, such as bees and butterflies, at a strawberry farm. This data can assist farmers in optimizing hive placement and numbers by enabling them to understand the contribution of each species to pollination. In 2016, Chen and Li created a robotic micro air vehicle that uses AI for autonomous pollination. The precise control of a robotic swarm is enabled by this small and efficient robot, which uses cameras and sensors to capture real-time data from crops and flowers.

- **Protected Cultivation**

Numerous advantages arise from integrating AI into protected horticulture, such as yield prediction, ensuring product quality from planting to harvest, optimizing resource allocation and time-to-market planning, and improving overall efficiency. The financial gains of growers and the long-term viability of their agricultural methods can both be greatly impacted by these developments. Artificial Intelligence is particularly important in large-scale greenhouse operations that are dispersed across multiple locations, where specialized labor is hard to come by and there is a growing demand for high-value food in metropolitan areas. Furthermore, it's critical to comprehend how growth circumstances and shelf-life procedures relate to one another. Adjustments to growing conditions can be informed by feedback from consumer acceptability, guaranteeing alignment with market expectations and improving total product quality.

- **Yield Forecasting**

Accurate enumeration of fruits is essential for the prediction of crop yields and the improvement of harvesting timetables to increase efficiency. Nonetheless, the manual tallying of fruits and vegetables is a laborious and time-consuming task. In response to this challenge, Lomte *et al.* (2019) presented an automated fruit counting mechanism that makes use of Image Processing technology. This mechanism analyzes input images obtained through a camera's sensor, identifying pertinent characteristics to produce the desired output data.

Maheswari *et al.* (2021) also talked about using Deep Learning (DL) techniques to build an advanced crop production estimation system. Their methodology consists of sampling trees, applying DL-based semantic segmentation frameworks to estimate yield, recognizing and counting fruits, supplementing data, and collecting data using a range of sensing instruments.

- **Aiding Harvesting**

Around 60% of production expenses is accounted for by manual harvesting of horticultural crops. The integration of AI into harvesting processes holds promise as a solution to labor shortages and high costs. The labor-intensive nature of picking can be significantly reduced with automated harvesters that move through rows of plants. Kitamura, S.*et al.* (2005) developed a fruit-picking robot that can differentiate between fruits and leaves using video image capture. With a camera attached, the robot arm recognizes colors and contrasts them with characteristics that are stored in memory. When a match is found, the fruit is carefully selected, applying just enough force to separate it from the tree without inflicting any harm. Birrell *et al.* (2019) came up with the 'Vegebot,' a robotic harvesting system that is tailored for iceberg lettuce, and it tackles issues like poor visibility and mechanical damage that are common in manual harvesting. Navas *et al.* (2021) studied how soft robotics and soft grippers made of silicone elastomers could be used to harvest delicate commodities. Soft grippers are thought to be suitable for high-value crops because they minimize mechanical damage and ensure the maximum market value of harvested products.

- **Processing Industry**

AI technology, ultrasonic sensing, and optical fluorescence imaging are all part of the self-optimizing clear-in-place system that is being developed in current research. The objective of this system is to discover tiny traces of food residue and microbial waste in food processing equipment. Enhancing cleaning efficiency, significantly reducing cleaning time, and minimizing resource consumption, especially water usage, can be achieved by doing so (Sahni *et al.,* 2021).

- **Quality Evaluation**

Assessing fruit quality during cold storage is crucial to ensure that high-quality fruits are delivered consistently to meet market demands. An Artificial

Neural Network (ANN) model was developed by Mohammed *et al.* (2022) to detect food safety and predict quality attributes in dates without requiring any invasive procedures. Their research demonstrated how well this method works to predict date fruit pH, TDS, water activity, and moisture content based on their electrical properties at 10 kHz. Sonwani *et al.* (2022) introduced a prototype that featured a range of sensors designed to identify and monitor food spoilage. The prototype consists of camera sensors and sensors for humidity, gas, and heat, which have the capability to notify users about food spoilage through voice commands or display notifications accurately at a rate of 95%.

- **Agrochemical Production**

Horticultural crops often depend on a variety of synthetic substances such as plant growth regulators, herbicides, and pesticides in order to achieve optimal yields and quality. Nevertheless, the presence of these substances in the soil and plants can lead to heightened levels of toxicity. Additionally, they have the potential to infiltrate the food chain, resulting in various disturbances at the metabolic level. Baruah (2019) has proposed that the integration of Machine Learning with Big Data Analytics could be employed to construct models for the production of suitable agrochemicals. These models would entail the utilization of pathogen, weed, and pest images to facilitate the formulation of effective solutions.

1.4 Challenges

Despite the enormous potential that artificial intelligence offers for applications in the field, most farmers in the nation are still unfamiliar with reliable high-tech technological solutions. Agriculture/horticulture is significantly impacted by external elements such as weather patterns and the presence of pests, which can be highly unpredictable. This variability can lead to inefficiencies in harvest planning due to fluctuations in these external variables. AI systems necessitate substantial volumes of data to train machines and provide accurate forecasts. While spatial data can be easily gathered for extensive agricultural areas, acquiring temporal data poses challenges due to various limitations. For many farmers in developing nations, especially those with small plots of land, financial resources are often limited, making it challenging to take on significant risks (Nayak *et al.* 2020). To address this, government assistance is crucial to make AI adoption feasible. The regulations pertaining to privacy, security, and ethics within the field of AI are frequently ambiguous and undergoing changes (Vijayakumar *et al.* 2021). It is crucial to conduct thorough testing and validation of emerging AI applications due to the significant impact of unpredictable environmental elements on farming activities. In order to guarantee the efficient utilization and ongoing improvement of these technologies, it is imperative to provide training to farmers on contemporary AI methodologies. Moving forward, efforts should be directed towards the integration of more sophisticated machinery, enhancing data collection practices, and optimizing software systems to better cater to the requirements of farmers.

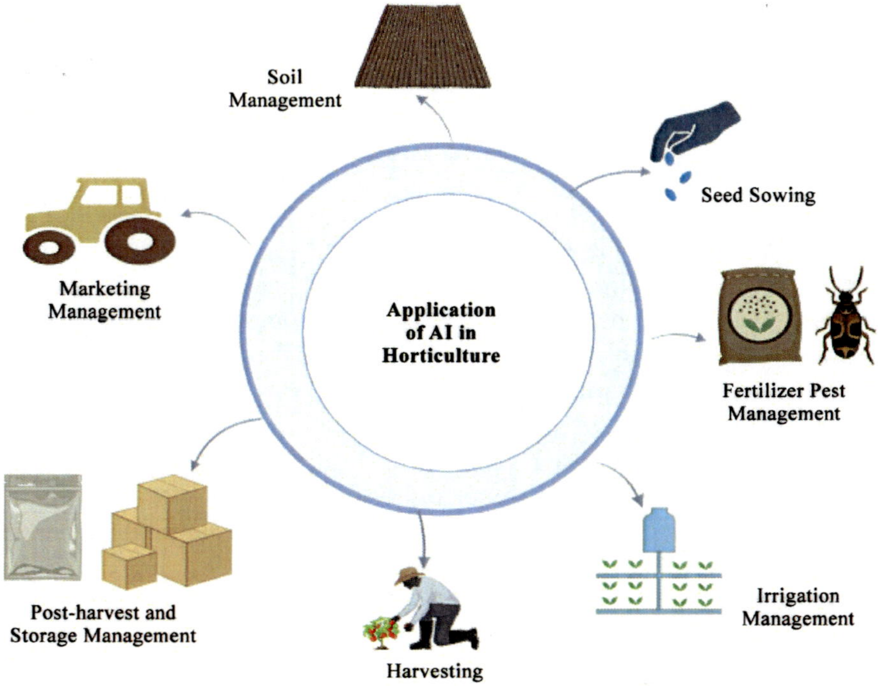

Fig. 1.2. Application of AI in horticulture

Table 1.1: Use of AI in different crops

Use of AI	Crop	Reference
Defect detection	Internal defect detection in Mango	Raghavendra*et al.* 2021
	Internal bruise detection in blueberries	Kuzy*et al.* 2018
Content detection	Soluble solids in Apple	Wang *et al.* 2022
	Soluble solids, starch pattern index, and Streif index in Apple	Cetin *et al.* 2022
Maturity and ripeness	Phillipine Coconut maturity grading	Caladcad*et al.* 2020
	Mango ripeness estimation	Wendel*et al.* 2018
	Cherry ripeness detection	Overbeck *et al.* 2017
Quality evaluation	Mass grading of Mangoes	Momin*et al.* 2017
	Infield grading and sorting system for Apples	Zhang *et al.* 2021

Harvester robot 1	Aubergines	Sepúlveda*et al.* 2020
Harvester robot 2 with LiDAR andencoder	Strawberry	Xiong*et al.* 2020
Harvester robot 3	Apple	Kang *et al.* 2020
Kubota farm vehicle	Grapes	Lowe *et al.* 2021
Robotic Shrimp for yield estimation	Apple	Bargoti*et al.* 2017
Pollinator robot	Kiwi	Williams *et al.* 2020
Bly-c-agri for Chemical weeding	Grapes	Sarri*et al.*2019

1.5 Conclusion

The way humans live is evolving due to artificial intelligence. With the dawn of artificial intelligence, there's a forecast that robots will gradually assume physical labor roles, while intelligent technology will take on mental challenges. It makes use of machine learning techniques and algorithms to give machines the ability to mimic particular cognitive processes. In agriculture, artificial intelligence not only helps farmers automate their processes but also helps it develop into precision cultivation, leading to enhanced crop yields and improved quality, all while reducing resource consumption. Even though artificial intelligence (AI) has a lot of potential for the agricultural industry, it's critical to recognize that adoption and effects will differ depending on location, the size of farm, and the technology available. Furthermore, we cannot ignore the moral questions raised by the use of AI on farms, including those pertaining to data privacy, data ownership, and employment implications.

References

Anonymous: Use of Artificial Intelligence in Agriculture. Press Information Bureau Government of India Ministry of Agriculture & Farmers Welfare. 2019; 17:44 IST.https://pib.gov.in/newsite/PrintRelease.aspx?relid=190957.

Bargoti, S. and Underwood, J.P., 2017. Image segmentation for fruit detection and yield estimation in Apple orchards. Journal of Field Robotics34 (6): 1039–1060.

Baruah, A. 2019.Artificial Intelligence in Indian Agriculture–An Industry and Startup Overview. Business Intelligence and Analytics.

Ben Ayed, R. and Hanana, M. 2021. Artificial intelligence to improve the food and agriculture sector. Journal of Food Quality 2021: 1-7.

Birrell, S., Hughes, J., Cai, J.Y. and Iida, F. 2019. A fieldtested robotic harvesting system for iceberg lettuce.Journal of Field Robotics37(2): 1-21.

Bryson, J.J. The artificial intelligence of the ethics of artificial intelligence: an introductory overview for law and regulation. In: Dubber, M.D., Pasquale, F., Das, S. (eds.) The Oxford Handbook of Ethics of AI.Oxford University Press; pp. 1–25.

Caladcad, J. A., Cabahug, S., Catamco, M. R., Villaceran, P. E., Cosgafa, L., Cabizares, K. N. and Hermosilla, M. 2020. Determining Philippine coconut maturity level using machine learning algorithms based on acoustic signal. Computers and electronics in agriculture 172: 105327.

Chaudhry S. andGarg S. 2019.Smart Irrigation Techniques for Water Resource Management. Smart Farming Technologies for Sustainable Agricultural Development: IGI Global. pp. 196–219.

Cheeti, S., Kumar, G. S., Priyanka, J. S., Firdous, G. and Ranjeeva, P. R. 2021.Pest detection and classification using YOLO AND CNN. Annals of the Romanian Society for Cell Biology15295-15300.

Chen, R. H. and Chen, C. C. 2022. Artificial Intelligence: An Introduction for the Inquisitive Reader. Chapman and Hall/CRC.

Chen, Y. and Li, Y. 2019. Intelligent autonomous pollination for future farming-a micro air vehicle conceptual framework with artificial intelligence and human-in-the-loop. IEEE Access 7: 119706-119717.

Çetin, N., Karaman, K., Kavuncuoğlu, E., Yıldırım, B., and Jahanbakhshi, A. 2022.Using hyperspectral imaging technology and machine learning algorithms for assessing internal quality parameters of apple fruits.Chemometrics and Intelligent Laboratory Systems230: 104650.

Fusco, R., Grassi, R., Granata, V., Setola, S. V., Grassi, F., Cozzi, D. andPetrillo, A. 2021. Artificial intelligence and COVID-19 using chest CT scan and chest X-ray images: machine learning and deep learning approaches for diagnosis and treatment. Journal of Personalized Medicine 11(10): 993.

Hafeez, A., Husain, M. A., Singh, S. P., Chauhan, A., Khan, M. T., Kumar, N. and Soni, S. K. 2022. Implementation of drone technology for farm monitoring & pesticide spraying: A review. Information processing in Agriculture10(2): 192-203.

Halpern, M. 2006. The trouble with the Turing test. The New Atlantis11: 42-63.

Kang, H., Zhou, H. andChen, C. 2020. Visual perception and modeling for autonomous apple harvesting. IEEE Access 8: 62151-62163.

Katrak, M. 2022. The role of language prediction models in contractual interpretation: the challenges and future prospects of GPT-3. Legal Analytics 47-62.

Kitamura, S., Oka, K. and Takeda, F. 2005.Development of picking robot in greenhouse horticulture. In: SICE Annual Conference (pp. 3176-3179).

Kodali, R.K. andSahu, A. 2016. An IoT based soil moisture monitoring on Losant platform. In: Proceedings of the 2nd International Conference on Contemporary Computing and Informatics, Greater Noida, India, 14–17 December 2016; IEEE: Piscataway, NJ, USA. pp. 764–768.

Kulkarni, V. A. and Deshmukh, A. G. 2013.Advanced agriculture robotic weed control system. International Journal of Advanced Research in Electrical, Electronics and Instrumentation Engineering2(10): 2320 – 3765.

Kumar, P.S. and Joshiba, G.J. 2019.Water Footprint of Agricultural Products. In: Muthu, S. (eds) Environmental Water Footprints. Environmental Footprints and Eco-design of Products and Processes.Springer, Singapore,pp. 1-19.

Kuzy, J., Jiang, Y. and Li, C. 2018. Blueberry bruise detection by pulsed thermographic imaging. Postharvest Biology and Technology 136: 166-177.

LeCun, Y., Bengio, Y. and Hinton, G. 2015. Deep learning. Nature 521(7553): 436-444.

Liao, S.M.A short introduction to the ethics of AI. 2020. In: Liao, S.M. (ed.) Ethics of Artificial Intelligence. New York, NY: Oxford University Press. pp. 1–42.

Liu, J. and Wang, X. 2020. Tomato diseases and pests detection based on improved Yolo V3 convolutional neural network. Frontiers in plant science 11: 521544.

Liu, S. Y. 2020. Artificial intelligence (AI) in agriculture. IT professional 22(3): 14-15.

Lomte, V.M., Sabale, S., Shirgaonkar, R., Nagathan, P. and Pawar, P. 2019. Fruit counting and maturity detection using image processing: A survey.International Journal of Research in Engineering Science and Management2(2): 809-812.

Lowe, T., Moghadam, P., Edwards, E. and Williams, J. 2021. Canopy density estimation in perennial horticulture crops using 3D spinning lidar SLAM. Journal of Field Robotics 38(4): 598-618.

Maheswari, P., Raja, P., Apolo-Apolo, O.E. and PerezRuiz, M. 2021. Intelligent fruit yield estimation for orchards using deep learning based semantic segmentation techniques—a review. Frontiers in Plant Science12: 1-18.

Marchetti, C. F., Ugena, L., Humplík, J. F., Polák, M., ĆavarZeljković, S., Podlešáková, K., ..andSpíchal, L. 2019. A novel image-based screening method to study water-deficit response and recovery of barley populations using canopy dynamics phenotyping and simple metabolite profiling. Frontiers in Plant Science 10: 445018.

McCarthy, J.,Minsky, M. L., Rochester, N. and Shannon, C. E. 2006. A proposal for the dartmouth summer research project on artificial intelligence, august 31, 1955. AI magazine 27(4): 12-12.

Mohammed, M., Munir, M. and Aljabr, A. 2022. Prediction of date fruit quality attributes during cold storage based on their electrical properties using artificial neural networks models. Foods11(11): 1-23.

Mohr, S. andKühl, R. 2021. Acceptance of artificial intelligence in German agriculture: an application of the technology acceptance model and the theory of planned behavior. Precision Agriculture 22(6): 1816-1844.

Momin, M. A., Rahman, M. T., Sultana, M. S., Igathinathane, C., Ziauddin, A. T. M. and Grift, T. E. 2017.Geometry-based mass grading of mango fruits using image processing. Information processing in agriculture 4(2): 150-160.

Navas, E., Fernandez, R., Sepulveda, D., Armada, M. and Gonzalez-de-Santos, P. 2021. Soft grippers for automatic crop harvesting: A review. Sensors 21(8): 1-27.

Nayak A K, Chatterjee D, Tripathi R, Shahid M, Vijayakumar S, Satapathy B S, Kumar A, Mohanty S, Bhattacharyya P, Mishra P and Kumar U. 2020. Climate smart agricultural technologies for rice production system in Odisha, pp. 366. ICAR-National Rice Research Institute, Cuttack, Odisha.

Overbeck, V., Schmitz, M., & Blanke, M. 2017.Non-destructive sensor-based prediction of maturity & optimum harvest date of sweet cherry fruit. Sensors (Switzerland) 17(2):277.

Pandey, D. K. and Chaudhary, B. 2016. Domestication-driven Gossypiumprofilin 1 (GhPRF1) gene transduces early flowering phenotype in tobacco by spatial alteration of apical/ floral-meristem related gene expression. BMC plant biology 16: 1-21.

Pandey, D. K. and Chaudhary, B. 2021. Transcriptional loss of domestication-driven cytoskeletal GhPRF1 gene causes defective floral and fiber development in cotton (Gossypium). Plant Molecular Biology 107: 519-532.

Pandya, R., Nadiadwala, S., Shah, R. and Shah, M. 2020.Buildout of methodology for meticulous diagnosis of K-complex in EEG for aiding the detection of Alzheimer's by artificial intelligence. Augmented Human Research 5: 1-8.

Patel, D., Shah, D. and Shah, M. 2020. The intertwine of brain and body: a quantitative analysis on how big data influences the system of sports. Annals of Data Science 7(1): 1-16.

Patel, D., Shah, Y., Thakkar, N., Shah, K. and Shah, M. 2020b. Implementation of artificial intelligence techniques for cancer detection. Augmented Human Research 5: 1-10.

Raghavendra, A., Guru, D. S., and Rao, M. K. 2021. Mango internal defect detection based on optimal wavelength selection method using NIR spectroscopy. Artificial Intelligence in Agriculture5: 43–51.

Sahni, V., Srivastava, S. and Khan, R. 2021.Modelling techniques to improve the quality of food using artificial intelligence.Journal of Food Quality 2021: 1-10.

Sarri, D., Martelloni, L., Rimediotti, M., Lisci, R., Lombardo, S. and Vieri, M. 2019. Testing a multi-rotor unmanned aerial vehicle for spray application in high slope terraced vineyard. Journal of Agricultural Engineering 50: 38-47.

SepúLveda, D., Fernández, R., Navas, E., Armada, M. & González-De-Santos, P. 2020. Robotic aubergine harvesting using dual-arm manipulation. IEEE Access 8: 121889-121904.

Shadrin, D., Menshchikov, A., Ermilov, D. and Somov, A. 2019.Designing future precision agriculture: Detection of seeds germination using artificial intelligence on a low-power embedded system. IEEE Sensors Journal 19(23): 11573-11582.

Sharma, R.. Artificial intelligence in agriculture: a review. 2021. In: 5th international conference on intelligent computing and control systems (ICICCS) (pp. 937-942). IEEE.

Sonwani, E., Bansal, U., Alroobaea, R., Baqasah, A. M. and Hedabou, M. 2022.An artificial intelligence approach toward food spoilage detection and analysis.Frontiers in Public Health 9: 1-13.

Stahl, B. C. 2021a. Artificial intelligence for a better future: an ecosystem perspective on the ethics of AI and emerging digital technologies. Springer Nature, pp. 124.

Stahl, B. C. 2021b. Ethical Issues of AI. In: B. C. Stahl (Ed.), Artificial Intelligence for a BetterFuture: An Ecosystem Perspective on theEthics of AI and Emerging Digital Technologies. Springer International Publishing, pp. 35-53.

Subeesh, A. and Mehta, C. R. 2021.Automation and digitization of agriculture using artificial intelligence and internet of things. Artificial Intelligence in Agriculture 5: 278-291.

Sukhadia, A., Upadhyay, K., Gundeti, M., Shah, S. & Shah, M. 2020. Optimization of smart traffic governance system using artificial intelligence. Augmented Human Research 5(1):13.

Talaviya, T., Shah, D., Patel, N., Yagnik, H. and Shah, M. 2020.Implementation of artificial intelligence in agriculture for optimisation of irrigation and application of pesticides and herbicides. Artificial Intelligence in Agriculture 4: 58-73.

Tzimas, T. 2021. Legal and ethical challenges of artificial intelligence from an international law perspective (Vol. 46). Springer Nature: Cham, Switzerland, pp. 9–32.

Vijayakumar, S., Subramanian, E., Saravanane, P., Gobinath, R. and Saha, S. 2021. Farm mechanisation in rice cultivation Present status, bottlenecks and potential. Indian Farming71(4).

Vijayakumar, V. and Balakrishnan, N. 2021. Artificial intelligence-based agriculture automated monitoring systems using WSN. Journal of Ambient Intelligence and Humanized Computing 12(7): 8009-8016.

Wang, F., Zhao, C., Yang, H., Jiang, H., Li, L. and Yang, G. 2022.Non-destructive and in-site estimation of apple quality and maturity by hyperspectral imaging. Computers and Electronics in Agriculture 195: 106843.

Wendel, A., Underwood, J. and Walsh, K. 2018. Maturity estimation of mangoes using hyperspectral imaging from a ground based mobile platform. Computers and Electronics in Agriculture 155: 298-313.

Williams, H., Nejati, M., Hussein, S., Penhall, N., Lim, J. Y., Jones, M. H., ... and MacDonald, B. 2020. Autonomous pollination of individual kiwifruit flowers: Toward a robotic kiwifruit pollinator. Journal of Field Robotics 37(2): 246-262.

Xiong, Y., Ge, Y., Grimstad, L. and From, P. J. 2020. An autonomous strawberry-harvesting robot: Design, development, integration, and field evaluation. Journal of Field Robotics 37(2): 202-224.

Zhang, Z., Lu, Y. and Lu, R. 2021. Development and evaluation of an apple infield grading and sorting system. Postharvest Biology and Technology 180: 111588.

2

Crop and Soil Monitoring in Horticulture Crops

Neha Sharma[1], Kishor Varotariya[2], Bismat Un Nisa[3]
Ume Kulsum[4] and Arizoo[5]

[1]*Division of Vegetable Science, Indian Agricultural Research Institute New Delhi- 110012, India*

[2]*Division of Vegetable Science, Indian Institute of Horticulture, Bengaluru-560089 India*

[3]*Division of Entomology, Sher-e-Kashmir University of Agricultural Sciences and Technology of Kashmir, Shalimar, Srinagar-190025, Jammu and Kashmir, India*

[4]*Division of Agricultural Statistics, Sher-e-Kashmir University of Agricultural Sciences and Technology of Kashmir, Shalimar, Srinagar-190 025, Jammu and Kashmir, India*

[5]*Division of Vegetable Science, Sher-e-Kashmir University of Agricultural Sciences and Technology of Kashmir, Shalimar, Srinagar-190025, Jammu and Kashmir, India*

Abstract

The integration of Artificial Intelligence (AI) in horticulture has revolutionized crop and soil monitoring techniques, enhancing productivity, sustainability, and resource management. This chapter explores the innovative applications of AI in the field of horticulture, focusing specifically on crop and soil monitoring. It discusses various AI techniques, including machine learning, computer vision, and remote sensing, and their utilization in monitoring crop growth, detecting diseases, optimizing irrigation, and assessing soil health. Through a comprehensive review of recent advancements, this chapter aims to provide insights into the transformative potential of AI in enhancing agricultural practices and ensuring food and nutritional security in horticultural systems.

2.1 Introduction

Horticulture plays a pivotal role in global food production and is essential for ensuring food security and improving nutrition. However, the increasing pressure on agricultural resources, coupled with environmental challenges, necessitates the adoption of innovative technologies to enhance productivity and sustainability. Artificial Intelligence (AI) has emerged as a powerful tool in this regard, offering unprecedented capabilities in data analysis, pattern recognition, and decision-making. In horticulture, AI holds immense potential for revolutionizing crop and soil monitoring practices, enabling more efficient resource utilization and better management strategies.

2.2 Foundations of Ai in Crop and Soil Monitoring

Artificial Intelligence (AI) forms the backbone of modern crop monitoring techniques, enabling the processing of vast amounts of agricultural data and the extraction of actionable insights. This section provides an in-depth exploration of the fundamental principles of AI, machine learning, and deep learning, elucidating their applications in crop monitoring within horticulture.

2.1.1 Understanding Artificial Intelligence (AI)

Definition and Scope: AI refers to the development of computer systems capable of performing tasks that typically require human intelligence, such as learning, reasoning, and problem-solving.

Types of AI: AI can be categorized into two broad types: narrow AI, which is designed for specific tasks, and general AI, which exhibits human-like intelligence across a wide range of domains. In the context of crop monitoring, narrow AI applications are more prevalent, focusing on tasks such as image recognition, predictive modeling, and decision support.

2.1.2 Machine Learning (ML) in Monitoring

Introduction to Machine Learning: Machine learning is a subset of AI that enables computers to learn from data and make predictions or decisions without being explicitly programmed. There are two types of machine learning viz- Supervised Learning: In supervised learning, algorithms learn from labeled training data, where each input is associated with a corresponding output or target variable. Supervised learning algorithms are widely used in crop monitoring for tasks such as yield prediction, disease classification, and pest detection.

Unsupervised Learning: Unsupervised learning involves training algorithms on unlabeled data to uncover hidden patterns or structures. Clustering algorithms, such as k-means and hierarchical clustering, can be applied to crop monitoring data to identify spatial or temporal trends, group similar crop instances, or detect anomalies.

2.1.3 Deep Learning for Complex Pattern Recognition

Introduction to Deep Learning: Deep learning is a subset of machine learning that employs neural networks with multiple layers to model complex relationships in data.

Neural Network Architecture: Deep learning models, inspired by the structure of the human brain, consist of interconnected layers of neurons. Common architectures include feedforward neural networks, convolutional neural networks (CNNs) for image processing, recurrent neural networks (RNNs) for sequential data, and transformers for natural language processing.

Applications in Crop Monitoring: Deep learning has revolutionized crop monitoring by enabling accurate and efficient analysis of large-scale agricultural datasets. CNNs, in particular, are well-suited for tasks such as plant disease detection, leaf segmentation, and crop classification based on aerial imagery or drone footage.

2.1.4 Data Preprocessing and Feature Engineering

Data Collection and Preparation: Data collection is a critical step in AI-driven crop monitoring, involving the acquisition of diverse datasets encompassing environmental variables, crop characteristics, and management practices. Data preprocessing techniques, such as cleaning, normalization, and imputation, are employed to ensure data quality and consistency.

Feature Selection and Extraction: Feature engineering involves selecting relevant input variables (features) that capture important information for the prediction or classification task at hand. Domain knowledge and statistical techniques are used to identify informative features, reduce dimensionality, and improve model performance.

2.2.5 Model Validation and Evaluation

Cross-Validation Techniques: Model validation is essential for assessing the performance of AI models and ensuring their generalization to unseen data. Cross-validation techniques, such as k-fold cross-validation and holdout validation, are commonly employed to evaluate model accuracy, precision, recall, and other performance metrics.

Hyperparameter Tuning: Hyperparameters are parameters that govern the learning process of AI models, such as the learning rate, regularization strength, and network architecture. Hyperparameter tuning techniques, including grid search and random search, are used to optimize model performance and prevent overfitting or underfitting.

2.1.6 Challenges and Considerations

Data Quality and Availability: One of the primary challenges in AI-driven crop monitoring is the availability and quality of training data. Limited access to labeled datasets, as well as issues related to data heterogeneity and bias, can affect the performance and reliability of AI models.

Interpretability and Transparency: Another consideration is the interpretability and transparency of AI models, particularly in decision-making processes with real-world implications. Ensuring that AI-driven recommendations are understandable and explainable to end-users is crucial for building trust and acceptance (Bodor *et al.*, 2023).

2.3 Crop And Soil Monitoring Using Ai

2.3.1 Crop Growth Prediction

Machine learning (ML) techniques have emerged as powerful tools for predicting various crop growth parameters in horticulture. These techniques utilize historical data on crop growth, environmental conditions, and management practices to develop predictive models that can forecast future outcomes. This section provides a detailed description of the application of machine learning in crop growth prediction, including the types of models used, data requirements, and case studies demonstrating their efficacy.

2.3.1.1 Introduction to Machine Learning in Crop Growth Prediction

Machine learning algorithms are capable of learning patterns and relationships from data to make predictions or decisions without being explicitly programmed. In the context of crop growth prediction, machine learning models can analyze historical data on factors such as temperature, precipitation, soil moisture, and nutrient levels to forecast key growth parameters, including yield, biomass, and phenological stages.

2.3.1.2 Types of Machine Learning Models

Several types of machine learning models are employed for crop growth prediction, each with its strengths and weaknesses:

Regression Models: Regression models, such as linear regression and polynomial regression, are used to predict continuous variables, such as crop yield or biomass, based on input features such as temperature, rainfall, and fertilizer application.

Decision Trees: Decision trees are hierarchical tree-like structures that recursively split the data based on feature values to make predictions. Decision tree algorithms, such as Random Forest and Gradient Boosting, are popular for their ability to handle non-linear relationships and interactions between variables.

Support Vector Machines (SVM): SVM is a supervised learning algorithm that is effective for both classification and regression tasks. SVM constructs a hyperplane in a high-dimensional space to separate data points into different classes or predict continuous outcomes.

Neural Networks: Neural networks, particularly deep learning architectures such as feedforward neural networks and recurrent neural networks (RNNs), have shown promise in predicting complex crop growth patterns from spatio-temporal data, such as satellite imagery and weather forecasts.

2.3.1.3 Data Requirements and Preprocessing

To train machine learning models for crop growth prediction, high-quality and diverse datasets are essential. These datasets typically include historical records of crop yields, environmental variables (e.g., temperature, precipitation), soil characteristics, and agronomic practices. Data preprocessing techniques, such as normalization, scaling, and handling missing values, are applied to ensure the quality and consistency of the data before model training.

2.3.1.4 Model Training and Evaluation

Machine learning models are trained using historical data, where the input features are used to predict the target variable (e.g., crop yield). The dataset is divided into training and testing sets, with the training set used to fit the model parameters and the testing set used to evaluate model performance. Various evaluation metrics, such as mean squared error (MSE), root mean squared error (RMSE), and coefficient of determination (R^2), are employed to assess the accuracy and generalization capability of the models (Huang *et al.*, 2022).

2.3.2 Case Studies and Applications

Several case studies demonstrate the application of machine learning in crop growth prediction across different horticultural crops and regions:

Forecasting Fruit Ripening: Machine learning algorithms were employed to predict the ripening time of fruits, such as tomatoes and bananas, based on environmental factors and physiological indicators. The models provided growers with insights into optimal harvesting times, reducing post-harvest losses and improving fruit quality (Aaron, 2020).

Predictive Analytics for Grapevine Growth: Vineyard managers utilized machine learning models to predict grapevine growth stages and optimize vineyard management practices, such as pruning, irrigation, and canopy management. The models leveraged historical data on weather conditions, soil moisture, and vine phenology to generate accurate growth forecasts throughout the growing season (Andrade, *et al.*, 2023).

Machine learning techniques offer a powerful means of predicting crop growth parameters in horticulture, enabling growers to optimize management practices, enhance productivity, and mitigate risks associated with environmental variability. By leveraging historical data and advanced modeling algorithms, machine learning facilitates data-driven decision-making in crop production, contributing to sustainable and resilient agricultural systems. Continued research and innovation in machine learning are essential for further advancing crop growth prediction capabilities and addressing emerging challenges in horticulture.

2.3.3 Integration of AI with Sensor Networks

Wireless sensor networks play a crucial role in collecting real-time data on environmental conditions, soil parameters, and crop status in horticultural systems. The integration of Artificial Intelligence (AI) with sensor networks enables advanced data analysis, pattern recognition, and decision-making, leading to optimized resource management and improved crop productivity. This section provides a detailed description of how AI is integrated with sensor networks in horticulture, covering the types of sensors used, data processing techniques, and practical applications.

2.3.3.1 Types of Sensors in Horticulture

Sensor networks in horticulture consist of various types of sensors deployed in the field to monitor different parameters:

Soil Sensors: Soil sensors measure parameters such as moisture content, temperature, pH level, and nutrient concentrations. Capacitive and resistive sensors are commonly used for soil moisture measurement, while pH and nutrient sensors utilize ion-selective electrodes or optical methods.

Weather Sensors: Weather sensors record meteorological data, including temperature, humidity, rainfall, wind speed, and solar radiation. These sensors provide valuable information for predicting weather patterns, assessing crop water requirements, and mitigating risks associated with extreme weather events.

Plant Sensors: Plant sensors monitor physiological parameters such as leaf temperature, chlorophyll content, and photosynthetic activity. Spectral sensors, thermal cameras, and fluorometers are used to measure plant responses to environmental stressors and assess crop health and vigor.

Remote Sensing Platforms: Remote sensing technologies, including satellites and unmanned aerial vehicles (UAVs), capture multispectral and hyperspectral imagery of agricultural landscapes. These platforms provide high-resolution spatial data for monitoring crop growth, detecting diseases, and assessing vegetation indices such as NDVI (Normalized Difference Vegetation Index) (Prabha *et al.*, 2023).

	Types of sensors	Purpose	
1.	Soil sensors	Estimate moisture content, temperature, pH level, and nutrient concentrations	
2.	Weather Sensors	record meteorological data, including temperature, humidity, rainfall, wind speed, and solar radiation	
3.	Plant Sensors	monitor leaf temperature, chlorophyll content, and photosynthetic activity	

2.3.3.2 Integration with Sensor Networks

The integration of AI with sensor networks involves several key steps:

Data Acquisition: Sensor networks collect raw data from multiple sensors distributed across the field. Data acquisition systems capture measurements at regular intervals and transmit them to a central database or cloud platform for storage and processing.

Data Processing: AI algorithms preprocess sensor data to remove noise, correct errors, and normalize measurements. Data preprocessing techniques, such as filtering, smoothing, and interpolation, enhance the quality and reliability of the sensor data for subsequent analysis.

Feature Extraction: AI models extract relevant features from sensor data to characterize environmental conditions, crop status, and growth dynamics. Feature extraction techniques, such as statistical analysis, Fourier transform, and wavelet analysis, identify informative patterns and trends in the sensor data.

Model Training and Prediction: AI models, such as machine learning algorithms and deep neural networks, are trained using historical sensor data and corresponding ground truth observations. These models learn the relationships between sensor measurements and target variables, such as crop yield, disease incidence, or irrigation scheduling, to make accurate predictions and decisions in real-time.

Decision Support Systems: AI-driven decision support systems (DSS) integrate sensor data with agronomic knowledge, weather forecasts, and economic models to provide actionable insights and recommendations to farmers and growers. DSS help optimize irrigation scheduling, fertilizer application, pest management, and harvesting operations based on current field conditions and future projections.

2.3.3.3 Practical Applications and Benefits

The integration of AI with sensor networks offers several practical applications and benefits in horticulture listed below:

1. **Precision Irrigation**: AI-driven sensor networks enable precise irrigation scheduling based on soil moisture levels, crop water requirements, and weather forecasts. By optimizing water use efficiency and reducing water wastage, precision irrigation conserves water resources and enhances crop yields.
2. **Nutrient Management**: Soil sensors combined with AI algorithms facilitate targeted nutrient application, minimizing fertilizer use while maximizing nutrient uptake by crops. Nutrient management strategies based on real-time sensor data improve soil fertility, reduce nutrient runoff, and mitigate environmental pollution.
3. **Disease Detection and Monitoring**: Plant sensors and remote sensing platforms integrated with AI enable early detection and monitoring of plant diseases and pests. By analyzing spectral signatures and physiological responses, AI models identify disease outbreaks and prescribe timely interventions, such as pesticide application or crop rotation, to prevent yield losses. Predictive analytics is another tool that AI systems may use to anticipate agricultural yields and identify pest or disease infestations. AI algorithms are highly accurate in predicting future yields by evaluating previous data on crop growth, weather patterns, and other pertinent aspects. This makes it possible for farmers to better organize their business, allocate resources optimally, and guarantee a steady supply of food to satisfy consumer demand.
4. **Yield Prediction and Forecasting**: Machine learning models trained on sensor data accurately predict crop yields and growth trajectories throughout the growing season. Yield forecasting based on real-time sensor measurements helps farmers optimize harvest schedules, plan

marketing strategies, and manage post-harvest logistics efficiently. Optical sensors mounted on unmanned aerial vehicles (UAVs) or ground-based platforms provide real-time data on crop health and yield potential. AI algorithms process this data to generate yield maps and optimize harvesting operations.

The integration of AI with sensor networks revolutionizes crop monitoring and management in horticulture, empowering growers with real-time insights and decision support tools. By leveraging sensor data and advanced analytics, AI-driven solutions optimize resource allocation, mitigate risks, and enhance sustainability in agricultural systems. Continued research and innovation in AI and sensor technologies are essential for further advancing precision agriculture and addressing the evolving challenges of food security and environmental sustainability.

Moreover, crop photos or sensor data may be analyzed by AI-powered disease detection systems to find early warning indicators of illnesses or pests. Farmers can stop future harm by taking quick action, including crop rotation or targeted pesticide use, by identifying these problems before they spread. By lowering the need for chemical inputs and minimizing crop losses, this proactive strategy promotes ecologically responsible and sustainable agricultural methods.

2.4 Soil Monitoring Using Ai

2.4.1 Sensor Networks for Soil Health Assessment

Soil health and quality are critical factors in horticultural productivity. AI-driven soil monitoring technologies provide precise, real-time insights into soil conditions, enabling optimal management practices. This section provides a detailed explanation of how AI is used for soil monitoring, the types of sensors involved, data processing techniques, and real-world case studies demonstrating successful applications.

Wireless sensor networks equipped with various sensors, including moisture, pH, and nutrient sensors, continuously monitor soil conditions. AI algorithms analyze sensor data to assess soil fertility, salinity levels, and nutrient deficiencies (Gautam & Pragya 2020; Millan *et al.,* 2020, Chiraz *et al*, 2022).

2.4.1.1 Types of Soil Sensors

Soil Moisture Sensors: Measure the volumetric water content in soil using methods such as time-domain reflectometry (TDR), capacitance, or resistive sensing.

Soil Temperature Sensors: Thermistors or thermocouples measure soil temperature, which affects microbial activity and nutrient availability.

Soil pH Sensors: Ion-selective electrodes measure the acidity or alkalinity of the soil, crucial for nutrient availability.

Nutrient Sensors: Detect concentrations of essential nutrients like nitrogen, phosphorus, and potassium using ion-selective electrodes or optical sensors.

Salinity Sensors: Measure the electrical conductivity of the soil to assess salinity levels, which can impact plant growth.

2.4.1.2 Data Processing Techniques

Data Preprocessing: Involves cleaning the data to remove noise and correct errors. Techniques include filtering, smoothing, and normalization.

Feature Extraction: Identifies key variables from the raw sensor data that are relevant for monitoring soil health. This could involve statistical measures, spectral analysis, or other methods.

Machine Learning Algorithms: Algorithms like regression models, decision trees, and neural networks analyze the processed data to detect patterns and make predictions about soil conditions.

Integration with GIS: Geographic Information Systems (GIS) help visualize and analyze spatial patterns in soil data across large agricultural fields.

2.4.1.3 Applications of AI in Soil Monitoring

1. Real-Time Monitoring: AI algorithms process continuous data streams from soil sensors to provide real-time insights into soil conditions, enabling immediate corrective actions.
2. Predictive Modeling: Predictive models forecast future soil conditions based on current data and historical trends, helping in planning irrigation, fertilization, and other management practices.
3. Decision Support Systems (DSS): DSS integrate soil data with other agronomic information to provide recommendations for optimizing soil health and crop productivity.

2.4.1.4 Case Studies

Case Study 1: Precision Agriculture in Vineyards (Johnson, 2003)

Location: Napa Valley, California

Objective: To optimize irrigation and improve grape quality by monitoring soil moisture and nutrient levels.

Sensors Used: Soil moisture sensors, nutrient sensors.

AI Techniques: Machine learning models were developed to predict soil moisture levels based on data from sensors and weather forecasts.

Data Integration: Sensor data was integrated with a GIS to map soil moisture variability across the vineyard.

Outcomes:

Water Savings: The precision irrigation system reduced water usage by 20%.

Improved Yield: Grape quality improved due to optimized soil moisture levels, leading to higher yields and better wine quality.

Case Study 2: Smart Farming in Potato Cultivation (Cherlinka, 2024)

Location: Idaho, USA

Objective: To manage soil nutrients and improve potato yield through real-time monitoring and AI-based predictions.

Sensors Used: Soil pH sensors, nutrient sensors.

AI Techniques: Deep learning models were used to analyze nutrient data and predict optimal fertilization schedules.

Data Integration: Data from soil sensors was combined with satellite imagery to monitor nutrient levels and plant health.

Outcomes:

Increased Efficiency: Nutrient application was optimized, reducing fertilizer usage by 15%.

Higher Productivity: Potato yields increased by 10%, and the incidence of nutrient-related diseases decreased.

AI-driven soil monitoring technologies significantly enhance the precision and efficiency of soil management in horticulture. By leveraging advanced sensors, data processing techniques, and predictive models, these systems provide real-time insights and actionable recommendations, leading to improved crop productivity and sustainability. The case studies highlight the diverse applications and benefits of AI in soil monitoring, demonstrating its potential to revolutionize horticultural practices globally.

2.4.2. Satellite Imagery for Soil Moisture Mapping

Satellite imagery has become a crucial tool in modern agriculture, particularly for soil moisture mapping. By providing extensive spatial coverage and frequent temporal updates, satellite-based soil moisture monitoring helps in effective water management, drought prediction, and precision agriculture. This section provides a detailed description of the methodologies, technologies, data processing techniques, and practical applications of using satellite imagery for soil moisture mapping, supported by case studies.

2.4.2.1 Importance of Soil Moisture Monitoring

Water Management: Essential for scheduling irrigation, preventing overwatering, and ensuring optimal plant growth.

Drought Prediction: Early detection of moisture deficits can help in mitigating drought impacts.

Crop Yield Forecasting: Soil moisture levels directly influence crop health and productivity.

2.4.2.2 Types of Satellites Used

S.no.	Types of Satellites Used	Use	
1.	Optical Satellites	Capture images in visible and infrared spectra, useful for assessing vegetation health and soil properties	
2.	Microwave Satellites	Active and passive microwave sensors are particularly effective for soil moisture estimation as they can penetrate the soil surface and provide data irrespective of weather conditions	
3.	Active Sensors	Send out a signal and measure the reflected backscatter (e.g., Synthetic Aperture Radar - SAR).	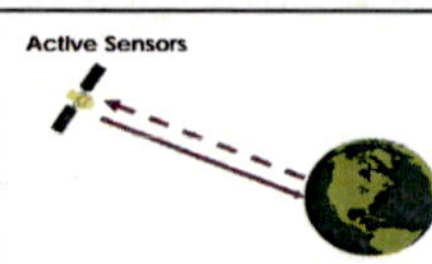
4.	Passive Sensors	Measure natural microwave emissions from the soil (e.g., Soil Moisture Active Passive - SMAP).	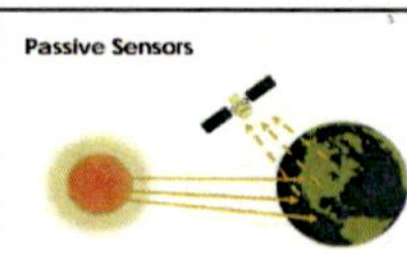

2.4.2.3 Methodologies for Soil Moisture Estimation

1. Optical Remote Sensing

Vegetation Indices: Indirectly estimate soil moisture by analyzing vegetation health using indices like NDVI (Normalized Difference Vegetation Index) and EVI (Enhanced Vegetation Index).

Thermal Infrared Imaging: Measures surface temperature, which can be correlated with soil moisture levels.

2. Microwave Remote Sensing

SAR (Synthetic Aperture Radar): Provides high-resolution soil moisture data by measuring the backscatter of radar signals from the soil surface.

SMAP (Soil Moisture Active Passive): Combines active and passive microwave sensing to provide global soil moisture maps with high accuracy.

2.4.2.4 Data Processing Techniques

1. Preprocessing

Radiometric Correction: Corrects sensor errors and calibration issues to ensure accurate reflectance values.

Geometric Correction: Aligns satellite images with geographical coordinates for accurate spatial analysis.

Noise Reduction: Filters out atmospheric and instrumental noise to improve data quality.

2. Soil Moisture Retrieval Algorithms

Empirical Models: Use statistical relationships between satellite measurements and ground truth soil moisture data.

Physical Models: Based on the physical properties of soil and its interaction with electromagnetic waves.

Machine Learning Models: Advanced algorithms (e.g., neural networks, support vector machines) trained on large datasets to predict soil moisture from satellite data.

3. Data Integration

Data Fusion: Combines data from multiple satellite sensors and platforms to improve soil moisture estimates.

GIS (Geographic Information Systems): Integrates satellite data with other geospatial datasets (e.g., topography, land use) for comprehensive soil moisture analysis.

2.4.2.5 Practical Applications and Benefits

1. Precision Agriculture

Irrigation Scheduling: Provides farmers with real-time soil moisture data to optimize irrigation practices and conserve water.

Fertilizer Application: Helps in determining the right timing and amount of fertilizer application based on soil moisture levels.

2. Drought Monitoring and Management

Early Warning Systems: Satellite data can be used to develop early warning systems for drought, allowing timely interventions.

Resource Allocation: Informs policymakers and resource managers for better allocation of water resources during drought conditions.

3. Environmental Monitoring

Flood Prediction: Soil moisture maps help in predicting flood risk areas by identifying regions with high soil saturation.

Soil Erosion Prevention: Monitors soil moisture to identify and manage areas at risk of erosion.

2.4.2.6 Case Studies

Case Study 1: Irrigation Management in California Vineyards(*Irrigation scheduling for viticulture using Sentinel-2*, October, 2021)

Objective: To optimize irrigation practices and improve water use efficiency in vineyards.

Satellites Used: Sentinel-1 (SAR), Landsat 8 (optical).

Techniques: Combined SAR and optical data to create high-resolution soil moisture maps.

Outcome: Reduced water usage by 25%, improved grape quality, and increased yield by optimizing irrigation based on soil moisture data.

Case Study 2: Drought Monitoring in Sub-Saharan Africa (Sheffield *et al.,* 2014)

Objective: To develop an early warning system for drought-prone regions.

Satellites Used: SMAP, MODIS (Moderate Resolution Imaging Spectroradiometer).

Techniques: Used SMAP data for soil moisture and MODIS for vegetation health to assess drought conditions.

Outcome: Provided timely drought alerts, enabling better preparedness and response, which reduced the impact on agriculture and food security.

Case Study 3: Flood Risk Assessment in the Netherlands (Markus *et al.*, 2010)

Objective: To predict and manage flood risks in low-lying areas.

SatellitesUsed: Sentinel-1 (SAR), SMAP.

Techniques: Integrated SAR data for soil moisture with hydrological models to predict flood-prone areas.

Outcome: Enhanced flood prediction accuracy, leading to better flood management and mitigation strategies.

Satellite imagery for soil moisture mapping is a transformative tool in modern horticulture, offering extensive spatial coverage and frequent updates. By integrating advanced remote sensing technologies with AI and GIS, these systems provide accurate, real-time soil moisture data, supporting precision agriculture, drought management, and environmental monitoring. The case studies illustrate the significant benefits and practical applications of satellite-based soil moisture mapping, demonstrating its potential to revolutionize agricultural practices and enhance sustainability.

2.5. Integration And Future Directions

The integration of crop and soil monitoring systems enabled by AI facilitates a holistic approach to horticultural management. By leveraging data-driven insights, growers can optimize input usage, minimize environmental impact, and maximize yields. However, several challenges, such as data interoperability, model interpretability, and scalability, need to be addressed to realize the full potential of AI in horticulture. Future research directions may involve the development of advanced AI algorithms capable of handling heterogeneous data sources, as well as the deployment of AI-enabled autonomous systems for real-time decision support on the farm.

2.6. Conclusion

In conclusion, the application of Artificial Intelligence in crop and soil monitoring has the potential to transform horticultural practices, leading to increased efficiency, sustainability, and resilience in agricultural systems. By harnessing the power of AI, growers can make informed decisions, optimize resource allocation, and mitigate risks associated with climate variability and pest outbreaks. Continued research and innovation in this field are crucial for unlocking the full benefits of AI-driven solutions and ensuring the future viability of horticulture in a rapidly changing world.

References

Aaron, D. 2020. Ripe Fruit Detection and Classification using Machine Learning. International Journal of Emerging Trends in Engineering Research. 8. 1845-1849. 10.30534/ijeter/2020/60852020.

Andrade, C.B., Moura-Bueno, J.M., Comin, J.J., Brunetto, G. 2023. Grape Yield Prediction Models: Approaching Different Machine Learning Algorithms. Horticulturae, 9, 1294. https://doi.org/10.3390/horticulturae9121294

Bodor, A., Hnida, M., & Daoudi, N. (2023). Machine Learning Models Monitoring in MLOps Context: Metrics and Tools. International Journal of Interactive Mobile Technologies (iJIM), 17(23), pp. 125–139.

Cherlinka, V., 2024.From soils to satellites': Harnessing EOSDA Crop Monitoring for sustainable potato cultivation in the U.S.potatonewstoday.com, USDAaccessed on 30 March 2024,https://www.potatonewstoday.com/2024/01/09/from-soils-to-satellites-harnessing-eosda-crop-monitoring-for-sustainable-potato-cultivation-in-the-u-s/.

Chiraz, M. C., Olfa, M. A. R. R. A. K. C. H. I., & Hamadi, H. A. B. A. I. E. B. (2022). Remote sensing and soil moisture data for water productivity determination. Agricultural Water Management, 263, 107482.

Gautam, D., & Pagay, V. (2020). A review of current and potential applications of remote sensing to study the water status of horticultural crops. Agronomy, 10(1), 140.

Huang, C. H., Chen, B. W., Lin, Y. J. and Zheng,J. X. "Smart Crop Growth Monitoring Based on System Adaptivity and Edge AI," in IEEE Access, vol. 10, pp. 64114-64125, 2022, doi: 10.1109/ACCESS.2022.3183277.

Johnson L.F., 2003. Temporal stability of an NDVI-LAIrelationship in a Napa Valley vineyard. Australian. Journal of Grape and Wine Research. 9(2):96-101.

Irrigation scheduling for viticulture using Sentinel-2, October, 2021, accessed on 15th May 2024 https://eo4society.esa.int/2021/10/06/irrigation-scheduling-for-viticulture-using-sentinel-2/

Millán, S., Campillo, C., Casadesús, J., Pérez-Rodríguez, J. M., & Prieto, M. H. (2020). Automatic irrigation scheduling on a hedgerow olive orchard using an algorithm of water balance readjusted with soil moisture sensors. Sensors, 20(9), 2526.

Markus, Adriaan & Courage, W.M.G. & Mierlo, M.. (2010). A Computational Framework for Flood Risk Assessment in The Netherlands. Scientific Programming. 18. 93-105. 10.3233/SPR-2010-0298.

Prabha,C., Malik,M., Kumari,S., Sharma,N., Sharma, A. and Khan,M. S. "A Review on Sensors and Technologies in Smart Farming Using AI and IoT Perspective and Their Challenges," 2023 IEEE 2nd International Conference on Industrial Electronics: Developments & Applications (ICIDeA), Imphal, India, 2023, pp. 322-327,doi:10.1109/ICIDeA59866.2023.10295251.

Sheffield, J., Wood, E. F., Chaney, N., Guan, K., Sadri, S., Yuan, X., Olang, L., Amani, A., Ali, A., Demuth, S., & Ogallo, L. (2014). A Drought Monitoring and Forecasting System for Sub-Sahara African Water Resources and Food Security. Bulletin of the American Meteorological Society, 95(6), 861-882. https://doi.org/10.1175/BAMS-D-12-00124.1

3

Robotics in Training and Pruning of Horticultural Crops

Kowsalya K.B[1]., Ramya Sree M.[2], Mahesh R.[3] and Talamarla Yeswanth Mahidar Gowd[4]

[1]Division of Vegetable Crops, ICAR-Indian Institute of Horticultural Research Bengaluru-560089, Karnataka

[2]Division of Fruits and Horticultural Technology, ICAR- Indian Agricultural Research Institute, New Delhi-110012

[3]Division of Agricultural Engineering, ICAR-Indian Agricultural Research Institute New Delhi-110012

[4]Dept of Vegetable science, College of Horticulture and forestry, Central Agricultural University, Pasighat-791102, Arunachal Pradesh

Abstract

Absolutely, the integration of robotics into horticulture presents unique challenges compared to industrial and commodity crop robotics. Horticultural crops, being living organisms, require specialized care and handling. The sensitivity of these crops to mechanical damage and environmental fluctuations adds another layer of complexity. Robots designed for horticultural tasks need to be versatile and adaptable to varying conditions. They must be capable of operating in environments where living plants are present, which requires precision and gentleness to avoid harming the crops. Additionally, these robots must navigate uneven terrain and cope with unpredictable weather conditions, which is quite different from the controlled environments of industrial settings. The classification of horticultural robots into different groups reflects the diverse tasks they need to perform, ranging from sowing and transplanting to harvesting and post-harvest handling. Each task presents its own set of challenges, requiring specialized robot designs and functionalities. The ultimate goal of creating robot farms, where all tasks are automated, is ambitious but achievable. However, it requires overcoming significant hurdles, such as ensuring robots can operate effectively in diverse and dynamic environments. This necessitates advancements in robotics technology, including improvements in sensors, actuators, and AI algorithms to enable robots to make real-time decisions and adapt to changing conditions. Indeed, the harmony between robotics and nature is crucial for the success of robot farms. Unlike factories, which are designed for uniformity and consistency, farms are inherently dynamic and subject to natural variability. Overcoming these challenges will require interdisciplinary collaboration among scientists, engineers, agronomists, and farmers to develop solutions that balance efficiency with sustainability and environmental stewardship.

3.1 Training

Virtual orchards are sophisticated, simulated environments used to train agricultural robots in a controlled, risk-free setting. These virtual landscapes replicate real-world orchards with high fidelity, incorporating detailed models of trees, fruits, branches, and various terrain features. By leveraging advanced technologies like 3D modelling, physics-based simulation, and machine learning, virtual orchards provide a comprehensive platform for developing and testing robotic algorithms without the logistical challenges and costs associated with field trials. In these simulations, agricultural robots can perform various tasks, such as navigating through rows of trees, detecting and localising fruits, and executing precise pruning actions (Shamshiri *et al.*, 2018). The virtual environment allows for testing different path planning and control algorithms, enabling researchers to fine-tune these systems for optimal performance.

Additionally, virtual orchards can simulate various environmental conditions, such as changing light levels, weather patterns, and the presence of obstacles, providing a robust testing ground for robots to adapt to dynamic scenarios. One significant advantage of virtual orchards is the ability to generate large amounts of training data for machine-learning models. Robots can be exposed to many pruning scenarios, enhancing their ability to generalize and perform effectively in real-world conditions. Furthermore, the use of virtual reality (VR) and augmented reality (AR) can aid in training human operators to control and supervise these robots, improving their proficiency and reducing the learning curve (Sarker, 2021).

Table 3.1 Summary of training methods and their results in fruit crops

Crop	**Training Method**	**Performance/Results**	**References**
Apple	Trellis-trained (Formal)	Increased productivity and profitability; simplified, compact canopy; good light penetration; high yield	Whiting (2018)

Sweet Cherry	Trellis-trained (Formal)	Machine harvest cost: $0.04/kg vs. $0.55/kg for manual; higher profitability but limited commercial adoption	Seavert and Whiting (2008)
Sweet Cherry	UFO	Facilitates automated field operations; angle to ground surface varies; similar benefits as formal apple trees	Peterson *et al.* (2003); He *et al.* (2015)
Sweet Cherry	Y-trellised (UFO/ Regular)	Higher fruit removal and recovery rates; reduced fruit damage; 60° angle from horizontal reduces drop height	Peterson *et al.* (2003); Ampatzidis and Whiting (2012)
Apple	Random Canopy	Maintains 2D fruiting-wall structure with pruning; offers similar accessibility for automated systems	Karkee *et al.* (2014)
Apple	Formal Canopy	Increasing opportunities to prevent fruiting on the trunk and instead maintaining it on branches, along with implementing targeted shake-and-catch harvesting techniques.	Hu *et al.* (2020); He (2018)

3.2 Pruning

The agricultural sector has experienced a notable enhancement in production efficiency through the adoption of mechanization and automation. Tree pruning is a crucial aspect of fruit tree orchard management that requires careful execution to balance immediate yield reduction with future yield improvement (Miller, 1960). This complex process involves informed decision-making to enhance tree health and productivity over the long term. Pruning serves various purposes, including reducing tree height to decrease harvesting costs, minimizing mutual shading, reducing inter-tree crowding, promoting regrowth, and mitigating risks.

Different pruning methods, such as hedging and selective limb removal, are effective in specific climatic conditions. The hedging is more suitable for warm subtropical areas, whereas limb removal benefits cool temperate regions (Menzel and Le Lagadec, 2014). Traditional pruning recommendations are often based on historical practices or studies of similar crops. The increasing scarcity of farm labour, particularly for time-sensitive and labour-intensive operations like pruning and training, poses a significant challenge for growers. To address these challenges, alternative methods such as mechanical (hedging) and robotic pruning are being explored. While mechanical pruning is effective, it tends to be non-selective and is often considered less preferable than precision pruning (Krueger *et al.*, 2013). On the contrary, robotic pruning provides a selective operation, presenting a potential solution to labor shortages (Lehnert, 2012).

Despite the primary focus of robotic systems on harvesting, their application in pruning encounters several challenges, such as the complex orchard environment and the necessity for precise and efficient technologies (Li *et al.*, 2011; Kapach *et al.*, 2012; Kootstra *et al.*, 2021). However, certain canopy architectures, like tall spindle, V-trellis, and fruiting walls, offer promising prospects for the advancement of robotic pruning systems due to their structures conducive to robotic operation (He and Schupp, 2018). Recent research endeavors have concentrated on developing algorithms for automated dormant tree pruning, which aids in shaping trees into desired configurations, thereby facilitating the adoption of robotic pruning systems (Kolmaniˇc *et al.*, 2021). To ensure the successful operation of a robotic pruning system, various components need integration, including sensing or perception, mechanical manipulation, and path planning.

3.3 Robotic Pruning

Robotic pruning entails the precise cutting of branches using a cutting tool affixed to a robotic arm (Lehnert, 2012). A typical robotic pruning system comprises a vision sensor, a mechanical manipulator, and an end-effector tool. The process commences with the detection and localization of branches via a camera vision system to pinpoint accurate pruning locations. Subsequently, the robotic arm maneuvers to reach the target point, ensuring proper orientation of the end-effector tool. Meanwhile, the end-effector tool executes the pruning cut. Although the angle of the pruning cut does not affect branch regrowth (Schupp *et al.*, 2019), the end-effector cutter must approach the branches at a specific orientation to perform the cut correctly, which is managed by path planning algorithms. Achieving the target pruning points involves several algorithms, including path planning (Zahid *et al.*, 2020), pruning cut sequencing, and obstacle avoidance (Botterill *et al.*, 2017).

3.4 Sensing or Perception

Identifying branches is a critical step in developing robotic pruners for tree fruit. It is essential for efficient path planning and obstacle avoidance. Machine vision, particularly through segmentation techniques, plays a pivotal role in detecting branches. Segmentation divides an image into distinct regions based on pixel characteristics, separating branches, leaves, fruits, and backgrounds into different classes. Various methods such as support vector machines (SVMs), convolutional neural networks (CNNs), region-based CNNs (R-CNNs), and mask R-CNNs have been extensively explored across different fruit orchards for effective branch detection.

Table 3.2 Summary of features and methods for branch detection

Crop	Features	Segmentation method	Accuracy rate	References
Apple	RGB, depth & pseudo-color images	R-CNN with AlexNet network	85.5%	Zhang *et al.* (2018)
	3D color images	Skeleton analysis by medial axis thinning algorithm	89 %	Karkee *et al.*, (2014)
	Color image	Contrast limited adaptive histogram equalization (CLAHE) based segmentation	94 %	Ji *et al.*, (2016)
Cherry	Color image	The Bayesian classifier to classify branch, cherry, leaf, and background	68 %	(Amatya and Karkee, 2016)
	RGB images	Bayesian classifier and morphological features	93.8 %	(Amatya *et al.*, 2016)
Citrus	RGB image	Multi-class support vector machine (SVM) and morphological operations	89.2 %	(Qiang *et al.*, 2014)
	Color and depth images	Mask R-CNN with a minimum enclosing rectangle (MER) bounding box	96.27 %	(Yang *et al.*, 2020)
Grape	RGB image	Linear Least Squares Error Fit	90 %	(Gao & Lu, 2006)
	Color image	Triangulation-feature-matching algorithm and Support vector machines (SVMs)	-	(Botterill *et al.*, 2017)
	Color image	Statistical pattern recognition algorithm based on graph morphometry	68 %	(Katyara *et al.*, 2020)
Litchi	Color and infrared images	Deeplabv3-based semantic segmentation, and 3D spatial positioning methods	83.33 %	(Li *et al.*, 2020)

(Zahid *et al.*, 2021)

3.5 Mechanical Manipulation

The manipulator's characteristics are primarily defined by its degrees of freedom (DoF) and types of joints (revolute or prismatic). Each joint in the manipulator contributes one degree of freedom, and the total DoFs and joint types significantly influence kinematic dexterity, obstacle avoidance capabilities, and spatial requirements during manipulation to achieve various positions and orientations of the end-effector tool (Burks *et al.*, 2018). Optimizing the manipulator's kinematics is challenging due to the inherent variability in tree architectures and workspace availability, underscoring the critical importance of selecting an appropriate joint configuration.

3.5.1 Open-loop Visual Control/Global Camera System: The system operates on hierarchical control principles that rely on precise 3D positioning, offering advantages such as enhanced stability in designated areas and straightforward operational control. However, open-loop visual control lacks the high accuracy necessary in vision systems, manipulator operation, and camera calibration. In grapevine pruning, this visual control method successfully achieved collision avoidance and completed the task in an execution time of 115 seconds per cane, with 85 percent of canes effectively positioned for cutting.

3.5.2 Closed-loop Feedback/ Visual Servo Control: The system operates through dynamic interaction between the manipulator and visual information, enabling object-friendly and real-time tracking without the need for calibration.

3.6 End Effectors

The end-effector plays a crucial role in the pruning robot, tasked with performing precise cuts on selected branches. Designing an effective end-effector tool for fruit tree pruning presents challenges due to crowded and overlapping branches, which limit maneuvering space. The design must account for mechanical aspects such as size, shape, weight, and maneuverability, as well as task-specific requirements related to physical, horticultural, and biological properties of the branches. Generally, end-effector tools for pruning or cutting can be categorized into two groups: saw discs or rotary cutters, and shear cutter devices.

Table 3.3 Summary of cutting mechanism and their performance

Crop	Cutting mechanism	Performance	References
Apple	Pneumatic shear cutting	Cut branches up to12mm in diameter	Zahid *et al.* (2020)
	Electric shear cutter	Cut the branches up to 25 mm diameter, 0.75 s per cut	Zahid *et al.* (2020)
Loquat	Electric disc saw	Cut the branches to 10 mm	Huang *et al.* (2016)
Grape	Electric mill end cutter	Cut 66 % of the canes, reached with 1 % of the ground truth	Botterill *et al.* (2017)
	Hydraulicshear cutter	Cut 85 % of the canes, 2.5 s per cut	Vision Robotics Corporation (2015)
Sweet cherry	Pneumatic shear cutter	92% cut success	You *et al.* (2020)

(Zahid *et al.*, 2021)

3.7 Path Planning

To enhance manipulation performance, various operations are conducted, including task sequencing, optimal trajectory planning, and obstacle avoidance, collectively known as path planning (Raja and Pugazhenthi, 2012). The predominant path planning strategy typically involves utilizing the manipulator's kinematic model to compute inverse kinematics for path establishment, often overlooking task sequencing and obstacle avoidance considerations (Hashimoto, 2003).

3.7.1 Algorithmic Approach

Algorithmic approaches in robotics for fruit crop training and pruning involve the use of advanced algorithms to analyze data, make decisions, and execute precise actions. These algorithms often incorporate machine learning, computer vision, and control systems to handle complex agricultural tasks.

Deep Learning Techniques for Path Prediction in Autonomous Pruning Robots

Deep learning techniques have revolutionized many fields, including robotics and precision agriculture. In the context of autonomous pruning robots, deep learning models can significantly enhance the efficiency and accuracy of path prediction and planning. Below is a generalised overview of how deep learning is applied to path prediction for these robots.

Key features	Function	Output
Deep Learning Models 1. Convolutional Neural Networks (CNNs) 2. Recurrent Neural Networks (RNNs) & Long Short-Term Memory Networks (LSTMs) 3. Generative Adversarial Networks (GANs)	Image recognition and processing	Identifying and localising fruits, branches, and obstacles in the environment
Training data	Large datasets of images and sensor data	Data includes labelled images of trees, branches, fruits, and obstacles
Feature Extraction	Extract relevant features	Branch density, size of the canopy etc
Path Prediction	Generating a sequence of movements	Avoid obstacles, reach target pruning areas, and optimise efficiency and effectiveness.
Real-time adaptation	Detect the obstacles	Recalibrate the path to avoid it

3.7.2 Simulation-based Approach

3.7.2.1 Tree/Branch Detection/Reconstruction using Machine Vision System

In automated pruning, the initial step involves detecting tree branches and pinpointing where cuts should be made on each branch. Effective sensing techniques are critical for identifying branches in fruit trees, selecting which branches to prune, and determining precise cutting points based on specific pruning guidelines. Once branches are detected, their locations can be mapped. To gather more detailed information about each branch, such as diameter, orientation, and overall tree structure, 3D reconstruction becomes necessary. Machine vision integrates sensors and algorithms to gather information about target objects and has been a staple in agricultural applications for decades (Chen *et al.*, 2002). A comprehensive review has outlined the sensors and technologies utilized in 3D imaging systems specifically designed for agricultural tasks (Vázquez-Arellano *et al.*, 2016). Various sensors, including cameras and LiDAR sensors, are employed in machine vision systems for detecting agricultural objects.

3.7.2.2 LiDAR-Scanned Models of Commercial Fruit Trees

A new framework has been developed for recommending pruning strategies using LiDAR-scanned models of commercial fruit trees. This framework focuses on enhancing light distribution within the canopy by utilizing a scoring function to evaluate the quality of tree shape and simulating the outcomes of pruning. Validation of this approach involves a three-stage process to ensure the framework's reliability and practicality.

Stage One: Scoring Function Development

In the initial stage, a scoring function is developed to assess the quality of tree shape by considering factors such as light availability and size, drawing from insights found in agricultural literature. This function is validated using actual yield data from avocado and mango orchards, showing a significant correlation with fruit count.

Stage Two: Pruning Simulation Tool

In the second stage, a pruning simulation tool is created. This tool uses algorithms to predict which parts of a tree's point cloud would be removed based on specified cut points, employing structural analysis techniques. The tool's precision is validated through experimental testing against manually created ground truth models of pruned trees, achieving promising results with an average F1 score of 0.78 across 144 experiments. This validation underscores the tool's capability in accurately forecasting the consequences of pruning decisions.

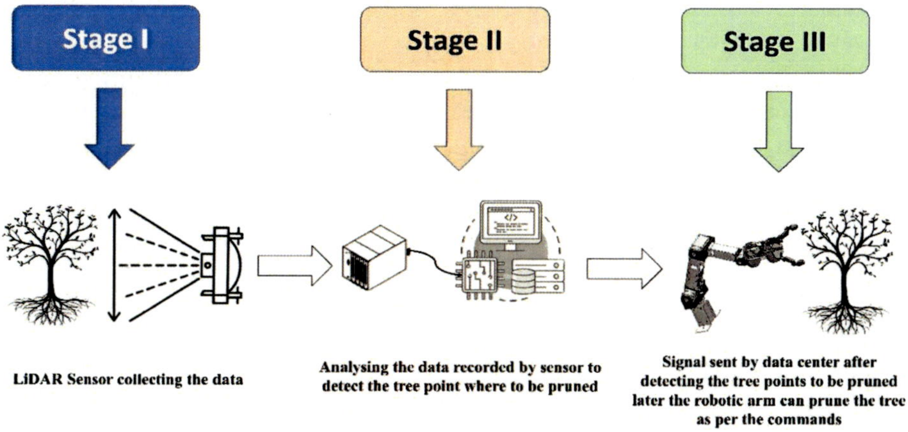

Fig 3.1 Stages of LiDAR-scanned models of commercial fruit trees

Stage Three: Pruning Suggestions and Evaluation

The final stage involves suggesting new pruning locations by identifying points in the tree that negatively impact light distribution. Using the previously developed scoring function and simulation tool, the framework estimates the improvement in light distribution after pruning. Additionally, certain cut points are identified that improve light distribution with minimal negative impact on tree volume.

3.7.3 Branch Detection/Reconstruction for Automated Pruning

Machine vision efforts in robotic pruning have predominantly concentrated on grapevines, benefiting from their relatively uniform and organized canopy structure. Researchers have employed various methods for branch detection in grapevines, such as single 2D cameras combined with image processing techniques to pinpoint pruning locations (Corbett-Davies *et al.*, 2012; Chattopadhyay *et al.*, 2016). Furthermore, a 3D machine vision system using stereo vision has been utilized to determine cutting points on grapevine branches by segmenting and measuring their lengths (Hosseini *et al.*, 2017). This system constructs a three-dimensional model of the vines, employs AI to decide which canes to prune, and utilizes a six-degree-of-freedom robotic arm for executing the cuts (Zhang *et al.*, 2017).

In contrast, advancements in robotic pruning for more complex canopies like apple trees have been notable. For example, a machine vision system utilizing a 3D camera has been proposed and evaluated for detecting and identifying branches in tall spindle apple trees. The developed algorithm relies on simple rules such as branch length and inter-branch spacing to determine pruning points. This approach effectively removed 85 percent of long branches and

69 percent of overlapping branches. Other studies have explored diverse vision-sensing technologies for automated pruning, including Kinect 2, RGBD sensors, depth imaging, and time-of-flight data. Tabb and colleagues have particularly focused on developing 3D reconstructions of fruit trees for automatic pruning, accurately identifying branch parameters like length, diameter, and angle (Tabb, 2017; 2018). The precision of branch detection and identification, the efficiency of branch reconstruction, and the system's cost are critical factors influencing the success of robotic pruning systems for tree branches.

LiDAR-scanned models of different commercial fruit trees

Crop	Model	Performance	References
Avocado	LiDAR-scanned models	Light distribution was improved by up to 25.15%	Westling *et al.* (2021)
Citrus	Unmanned aerial vehicle (UAV)-based light detection and ranging (LiDAR) technology	Fenced mode 71.94 %	Dian *et al.* (2023)
Sweet cherry	Terrestrial LiDAR	Detect differences in canopy growth habits	Zhang *et al.*, (2017)
Olive	Pruning of Unvibrating Localized Portions (PULP) System Approach	Precision pruning method for trees harvested with trunk shakers.	Vicino *et al.* (2020)
Grape	Stereo vision/3D vision	Accurate canopy volume determination	Hosseini and Jafari, (2017)
Apple	High-quality tractor-mounted LiDAR	Targeted root-pruning of vigorous trees	Siefen *et al.* (2023)
	LiDAR sensor	Relationship between canopy volume and number of leaves	Mahmud *et al.* (2021)
Pear	LiDAR-scanned models	Prediction of illumination based on different leaf-area characteristics	Tang *et al.* (2024)
	LiDAR-RTK fusion acquisition system	Construct 3D prescription maps autonomously that match the application requirements in digital orchard management.	Han *et al.* (2023)
Almond	LiDAR sensor	Expand the possibilities of using multispectral images in orchard management, particularly in hedgerow plantations.	Martínez-Casasnovas *et al.* (2021)

3.8 Conclusion

The application of robotics in horticulture differs markedly from its use in industrial and commodity crop contexts due to the living nature of horticultural produce, which requires gentle handling and precise environmental conditions. In horticulture, robots are categorized into various groups based on their functions, including sowing, transplanting, weeding, disease control, maintenance, harvesting, and post-harvest handling. A primary goal is the development of fully automated robot farms capable of performing all agricultural tasks, although the heterogeneous nature of farm environments poses significant challenges. Unlike factory robots operating in predictable settings, agricultural robots must navigate uneven terrain and adapt to constantly changing conditions.

Pruning, crucial in orchard management, aims to balance immediate yield reduction with long-term productivity gains. Traditional methods are increasingly supplemented by mechanical and robotic pruning to address labor shortages and enhance precision. Robotic pruning systems integrate multiple components, including sensing or perception, mechanical manipulation, and path planning. These systems leverage advanced machine vision techniques for branch detection, utilizing methods such as CNNs, R-CNNs, and SVMs to achieve high accuracy rates across diverse crops. Mechanical manipulation is facilitated by manipulators with optimized degrees of freedom and joint configurations for enhanced kinematic agility and obstacle avoidance. End-effectors, such as shear cutters and disc saws, are engineered to execute precise cuts in dense and overlapping branch environments. Deep learning techniques further improve path prediction and obstacle avoidance, enabling robots to adapt dynamically to changing environments in real-time. Simulation-based approaches, including virtual orchards, provide a risk-free environment for training and refining robotic algorithms. These simulations generate valuable data for machine learning models and enhance human operator proficiency through VR and AR technologies. Recent advancements in 3D reconstruction and LiDAR-scanned models have enabled the development of frameworks for recommending pruning strategies based on optimizing light distribution and tree shape quality. These frameworks incorporate scoring functions, pruning simulation tools, and algorithms for suggesting pruning actions to enhance light availability and optimize tree health and productivity.

In conclusion, the integration of advanced robotics and simulation technologies in horticulture shows promise for improving efficiency, precision, and sustainability in orchard management. While significant challenges remain, particularly in adapting robotic systems to the variability of natural environments, ongoing research and technological innovations continue to push the boundaries of automated horticulture. The future of farming may see the emergence of fully automated robot farms where machines work harmoniously with nature to meet the demands of modern agriculture.

References

Ampatzidis, Y. G., & Whiting, M. (2012, April). Comparing the efficiency of future harvest technologies for sweet cherry. In I International Symposium on Mechanical Harvesting and Handling Systems of Fruits and Nuts 965 (pp. 195-198). Doi: 10.17660/ActaHortic.2012.965.26

Botterill, T., Paulin, S., Green, R., Williams, S., Lin, J., Saxton, V., Mills, S., Chen, X.Q., & Corbett-Davies, S. (2017). A robot system for pruning grape vines. Journal of Field Robotics, 34(6), 1100-1122. Doi: 10.1002/rob.21680.

Burks, T. H. O. M. A. S., Bulanon, D. U. K. E., & Mehta, S. (2018). Opportunity of robotics in precision horticulture. Automation in tree fruit production: Principles and practice.

Chattopadhyay, S., Akbar, S. A., Elfiky, N. M., Medeiros, H., & Kak, A. (2016, March). Measuring and modeling apple trees using time-of-flight data for automation of dormant pruning applications. In 2016 IEEE Winter conference on applications of computer vision (WACV) (pp. 1-9). IEEE. Doi: 10.1109/WACV.2016.7477596.

Chen, Y. R., Chao, K., & Kim, M. S. (2002). Machine vision technology for agricultural applications. Computers and electronics in Agriculture, 36(2-3), 173-191. Doi: 10.1016/S0168-1699(02)00100-X

Corbett-Davies, S., Botterill, T., Green, R., & Saxton, V. (2012, November). An expert system for automatically pruning vines. In Proceedings of the 27th conference on image and vision computing New Zealand (pp. 55-60). Doi: 10.1145/2425836.2425849

Dian, Y., Liu, X., Hu, L., Zhang, J., Hu, C., Liu, Y., Zhang, J., Zhang, W., Hu, Q., Zhang, Y. & Fang, Y. (2023). Characteristics of photosynthesis and vertical canopy architecture of citrus trees under two labor-saving cultivation modes using unmanned aerial vehicle (UAV)-based LiDAR data in citrus orchards. Horticulture Research, 10(3), uhad018. Doi: 10.1093/hr/uhad018

Han, L., Wang, S., Wang, Z., Jin, L., & He, X. (2023). Method of 3D Voxel Prescription Map Construction in Digital Orchard Management Based on LiDAR-RTK Boarded on a UGV. Drones, 7(4), 242. Doi: 10.3390/drones7040242

He, L., & Schupp, J. (2018). Sensing and automation in pruning of apple trees: a review. Agronomy 8 (10), 211. Doi: 10.3390/agronomy8100211.

He, L., Zhou, J., Zhang, Q., &Karkee, M. (2015). Evaluation of multipass mechanical harvesting on 'Skeena'sweet cherries trained to Y-trellis. HortScience, 50(8), 1178-1182. Doi: 10.21273/HORTSCI.50.8.1178

Hosseini, S. M., & Jafari, A. (2017). Designing an algorithm for pruning grapevine based on 3D image processing. Iranian Journal of Biosystems Engineering, 48(3), 289-297. Doi: 10.22059/IJBSE.2017.205096.664772

Hu, G., Bu, L., & Chen, J. (2020). Simulation to determination of significant parameters on apple stress for combing harvesting in trellis trained trees. Scientia Horticulturae, 274, 109654. Doi: 10.1016/j.scienta.2020.109654

Kapach, K., Barnea, E., Mairon, R., Edan, Y., & Ben-Shahar, O. (2012). Computer vision for fruit harvesting robots–state of the art and challenges ahead. International Journal of Computational Vision and Robotics, 3(1-2), 4-34. Doi: 10.1504/IJCVR.2012.046419.

Karkee, M., Adhikari, B., Amatya, S., & Zhang, Q. (2014). Identification of pruning branches in tall spindle apple trees for automated pruning. Computers and Electronics in Agriculture, 103, 127-135. Doi: 10.1016/j.compag.2014.02.013

Kolmanič, S., Strnad, D., Kohek, Š., Benes, B., Hirst, P., &Žalik, B. (2021). An algorithm for automatic dormant tree pruning. Applied Soft Computing, 99, 106931. Doi: 10.1016/j. asoc.2020.106931.

Kootstra, G., Wang, X., Blok, P. M., Hemming, J., & Van Henten, E. (2021). Selective harvesting robotics: current research, trends, and future directions. Current Robotics Reports, 2, 95-104. Doi: 10.1007/s43154-020-00034-1.

Krueger, W. H., Niederholzer, F. J. A., & Fichtner, E. (2012, May). Investigation of pruning strategies for dried plums including hand, mechanical and combinations. In X International Symposium on Plum and Prune Genetics, Breeding and Pomology 985 (pp. 201-207). Doi: 10.17660/ActaHortic.2013.985.25.

Lehnert, R., 2012. Robotic pruning. Good Fruit Grower Nov. 1, 2012. https://www. goodfruit. com/robotic-pruning/.

Li, P., Lee, S. H., & Hsu, H. Y. (2011). Review on fruit harvesting method for potential use of automatic fruit harvesting systems. Procedia Engineering, 23, 351-366. Doi: 10.1016/j. proeng.2011.11.2514.

Mahmud, M. S., Zahid, A., He, L., Choi, D., Krawczyk, G., Zhu, H., & Heinemann, P. (2021). Development of a LiDAR-guided section-based tree canopy density measurement system for precision spray applications. Computers and Electronics in Agriculture, 182, 106053. Doi: 10.1016/j.compag.2021.106053

Martínez-Casasnovas, J. A., Sandonís-Pozo, L., Escolà, A., Arnó, J., & Llorens, J. (2021). Delineation of management zones in hedgerow almond orchards based on vegetation indices from UAV images validated by LiDAR-derived canopy parameters. Agronomy, 12(1), 102. Doi: 10.3390/agronomy12010102

Menzel, C. M., & Le Lagadec, M. D. (2014). Increasing the productivity of avocado orchards using high-density plantings: a review. Scientia Horticulturae, 177, 21-36. Doi: 10.1016/j.scienta.2014.07.013

Miller, M. (1960). Avocado pruning to regulate crop production. Yearbook of the California Avocado Society, 44, 42-4.

Peterson, D. L., Whiting, M. D., & Wolford, S. D. (2003). Fresh–market quality tree fruit harvester part I: sweet cherry. Applied Engineering in Agriculture, 19(5), 539. Doi: 10.13031/2013.15313

Sarker, I. H. (2021). Deep learning: a comprehensive overview on techniques, taxonomy, applications and research directions. SN Computer Science, 2(6), 420.doi: 10.1007/ s42979-021-00815-1

Schupp, J. R., Winzeler, H. E., & Schupp, M. A. (2019). Stub length and stub angle did not influence renewal shoot number or branch angle of tall spindle 'Gala'/Malling 9 apple trees. HortTechnology, 29(1), 46-49. Doi: 10.212 73/HORTTECH04218-18.

Seavert, C. F., & Whiting, M. D. (2008). Comparing the economics of mechanical and traditional sweet cherry harvest. In IX International Symposium on Integrating Canopy, Rootstock and Environmental Physiology in Orchard Systems 903 (pp. 725-730). Doi: 10.17660/ActaHortic.2011.903.101

Shamshiri, R., Hameed, I. A., Pitonakova, L., Weltzien, C., Balasundram, S. K., J Yule, I., & Chowdhary, G. (2018). Simulation software and virtual environments for acceleration of agricultural robotics: Features highlights and performance comparison.

Siefen, N., McCormick, R. J., Vogel, A. M., & Biegert, K. (2023). Effects of laser scanner quality and tractor speed to characterise apple tree canopies. Smart Agricultural Technology, 4, 100173. Doi: 10.1016/j.atech.2023.100173

Tabb, A., & Medeiros, H. (2017, September). A robotic vision system to measure tree traits. In 2017 IEEE/RSJ International Conference on Intelligent Robots and Systems (IROS) (pp. 6005-6012). IEEE.

Tabb, A., & Medeiros, H. (2018). Automatic segmentation of trees in dynamic outdoor environments. Computers in Industry, 98, 90-99. Doi: 10.1016/j.compind.2018.03.002

Tang, H., Xu, Z., Chen, S., Yang, S., Liu, W., Zhou, L., ... & Zheng, Y. (2024). Light efficacy estimation for fruit trees based on LIDAR point clouds: A case study on pear trees. Scientia Horticulturae, 324, 112590. Doi: 10.1016/j.scienta.2023.112590

Tombesi, S., Poni, S., Palliotti, A., & Farinelli, D. (2017). Mechanical vibration transmission and harvesting effectiveness is affected by the presence of branch suckers in olive trees. Biosystems Engineering, 158, 1-9. Doi: 10.1016/j.biosystemseng.2017.03.010

Vázquez-Arellano, M., Griepentrog, H.W., Reiser, D., & Paraforos, D.S. (2016). 3-D imaging systems for agricultural applications—a review. Sensors, 16(5): 618. Doi: 10.3390/s16050618

Westling, F., Underwood, J., & Bryson, M. (2021). A procedure for automated tree pruning suggestion using LiDAR scans of fruit trees. Computers and Electronics in Agriculture, 187, 106274. Doi: 10.1016/j.compag.2021.106274

White, N., & Hanan, J. (2014, August). A model of macadamia with application to pruning in orchards. In XXIX International Horticultural Congress on Horticulture: Sustaining Lives, Livelihoods and Landscapes (IHC2014): 1109 (pp. 75-82). Doi: 10.17660/ActaHortic.2016.1109.12

Whiting, M. A. T. T. H. E. W. (2018). Precision orchard systems. Automation in tree fruit production: Principles and practice, 75-93.

You, A., Parayil, N., Krishna, J. G., Bhattarai, U., Sapkota, R., Ahmed, D., ... & Davidson, J. R. (2022). An autonomous robot for pruning modern, planar fruit trees. arXiv preprint arXiv:2206.07201. Doi: 10.48550/arXiv.2206.07201

Zahid, A., He, L., Choi, D. D., Schupp, J., & Heinemann, P. (2020). Collision free path planning of a robotic manipulator for pruning apple trees. In 2020 ASABE annual international virtual meeting (p. 1). American Society of Agricultural and Biological Engineers. Doi: 10.13031/aim.202000439.

Zahid, A., Mahmud, M. S., He, L., Heinemann, P., Choi, D., & Schupp, J. (2021). Technological advancements towards developing a robotic pruner for apple trees: A review. Computers and Electronics in Agriculture, 189, 106383. Doi; 10.1016/j.compag.2021.106383

Zhang, J., He, L., Karkee, M., Zhang, Q., Zhang, X., & Gao, Z. (2017). Branch detection with apple trees trained in fruiting wall architecture using stereo vision and regions-convolutional neural network (R-CNN). In 2017 ASABE annual international meeting (p. 1). American Society of Agricultural and Biological Engineers. Doi: 10.13031/aim.201700427

4

Role of AI in Diagnosis and Monitoring of Diseases and Insect Pests

Gazala Gulzar* and Suhail Quyoom Wani

Division of Plant Pathology, Sher-e Kashmir University of Agricultural Sciences and Technology, Shalimar, Srinagar-190025, J&K, India

Abstract

For developing nations, the agricultural sector provides the backbone of growth in the economy. The biggest threats to crop output are pests and diseases, which can have a significant negative impact on the economy. Accurate and timely detection of pests and plant diseases is essential to attaining sustainable farming practices and long-term agricultural productivity gains. These dangerous organisms can damage the roots, stems, and leaves of plants, among other plant parts. Conventional plant disease detection techniques are time-consuming, resource-intensive, and require specific knowledge. For this reason, there is a growing trend toward automating plant diagnostics through the use of artificial intelligence (AI) in conjunction with Internet of Things (IoT) sensors. To put it briefly, AI techniques have the potential to revolutionize several aspects of plant protection. In order to address even more complex ecological questions, future research efforts should focus on improving AI tools and incorporating them into entomological and pathological studies with ease.

Keywords: *Plant disease, pest detection, artificial intelligence, internet of things, sustainable agriculture.*

4.1 Introduction

Pests and plant diseases pose serious risks to global agriculture (Strange and Scott, 2005; Oerke, 2006). Plant diseases and pests are the natural disasters that can impede a plant's normal growth throughout the plant's whole life cycle, from seed formation to seedling growth and may result in plant death (Liu and Wang, 2021). Diseases reduce yields in a number of important crops by 21–30% globally (Savary *et al.*, 2019), whereas insect infestations reduce crop output by 30–33% annually (Kumar *et al.*, 2019). For the management, accurate disease diagnosis and pest identification are essential. Artificial intelligence (AI) techniques can be a crucial addition to enable the monitoring of plant diseases and pests at a coarse scale, as traditional field scouting of these pests and diseases is labor-intensive, subject to bias, and typically shows low efficiency (Mahlein, 2016).

Agriculture is undergoing a revolution because of artificial intelligence (AI), which is replacing outdated techniques with more effective ones (Talaviya *et al.*, 2020). AI intervention in agriculture is helping farmers increase productivity and reduce adverse environmental effects (Sujawat, 2021). The application of AI techniques has created new opportunities for scientific research as well as new instruments and ways for handling and analyzing vast volumes of data. An important benefit of AI is its capacity to spot patterns in data that are not immediately obvious or available to humans. This allows for previously unattainable new insights and discoveries (Angermueller *et al.*, 2016; Eraslan *et al.*, 2019). AI is currently changing how we work and live, yet its influence on pathology and entomology research is still underappreciated. AI's cost-effectiveness, precision, quickness of performance, and adaptability are its core concepts in agriculture (Eli-Chukwu, 2019). In agriculture, artificial intelligence not only helps farmers use their abilities, but it also guides them to obtain better quality and larger yields with fewer resources (Khandelwal and Chavhan, 2019). AI-based technology addresses the issues that agricultural sector faces and helps to increase efficiency in all areas. These challenges include crop harvesting, irrigation, soil content sensitivity, crop monitoring, weeding, harvesting, and establishing.

AI sensors can detect and identify weeds, and the technology assists farmers in diagnosing plant diseases, pests, and malnourishment. For the recently developed Machine Learning (ML) and Deep Learning (DL), Artificial Intelligence can provide a workable and efficient solution (Murugesan *et al.*, 2019). We now have a clear method for detecting plant diseases on a massive scale by using machine learning to train the vast data sets that are publically available. Machine learning techniques are employed to identify and categorize diseases on agricultural goods, such as different types of plants, fruits, and vegetables. A robot that uses machine learning and image processing to detect leaf diseases is described (Kumar and Vani, 2018). Deep learning records information about past cases that may be useful in identifying plants with similar symptoms or diseases, in addition to assisting in the identification of important relationships within the data.

Conventional image processing algorithms or human feature creation along with classifiers are frequently employed for classic machine vision-based plant disease and pest detection methods (Tsaftaris *et al.*, 2016). When designing an imaging scheme, this type of technology typically takes into account the many characteristics of plant diseases and pests. It also selects the right light source and shooting angle, which helps produce photos with consistent illumination. Plant diseases and pests can be difficult to identify in real complex natural environments due to a variety of factors, including low contrast, huge scale changes between different types of lesions and the surrounding area, and high levels of noise in the lesion image. Additionally, gathering photos of pests and diseased crops under some circumstances causes a lot of disruptions. Better detection results are currently hard to come by and the old classical procedures frequently seem futile. Convolutional neural networks (CNNs) have been used successfully in deep learning models in many computer vision (CV) and computer-vision fields in recent years. Examples of these fields include Medical Image Recognition (Sundararajan *et al.*, 2019), Expression Recognition (Li *et al.*, 2019), Traffic Detection (Yang *et al.*, 2019), Facial Recognition (Kumar and Singh, 2020), Scenario Text Detection (Melnyk *et al.*, 2020), etc. Numerous deep learning-based techniques for detecting plant diseases and pests are used actual agriculture, and a range of deep learning-based software for detecting pests and plant diseases have been created. Consequently, the artificial intelligence-based approach for detecting plant diseases and pests has significant research value in academia and a wide range of potential commercial applications.

4.2 Principles in Detecting Plant Diseases And Insect Pests

Agrochemicals are one means of controlling diseases and insect pests, however using them raises production costs and has a detrimental effect on the environment. Because of this, integrated disease management (IDM) and integrated pest management (IPM) aim to lower hazards to human and animal health as well as the environmental impact of using pesticides and fungicides. The monitoring of diseases and insect pests, which are currently aided by sensing and artificial intelligence (AI) techniques, is a critical component of such a management system (Toscano-Miranda *et al.*, 2022). It should be emphasized, that not all plant diseases and pests can be detected remotely because some of them lack distinguishable features. This is important for researchers to bear consideration when studying the relationships between the pathogen and the host. On the other hand, certain root infections and soil-borne diseases that have a systemic impact on plant physiology can also be identified. Therefore, the presence of a particular response that can be picked up by a particular sensor system is a necessary condition for the detection and monitoring of plant diseases and pests using AI.

In addition to the physiological alterations that pests or diseases induce in plants, three different kinds of damages related to remote sensing have been identified. One includes the loss of LAI and reduction of biomass. This kind of damage typically happens as a result of pest infestations that consume plant parts, significantly reducing biomass and leaf area (Zhang *et al.*, 2015). Despite this, the monitoring of such deterioration is always afflicted by high uncertainty due to its lack of spectrum specificity. Second damage is due to infected lesions or pustules. The color and shape of lesions, also known as pustules, might differ throughout diseases and pests. It should be emphasized that these lesions and pustules in canopy distribution and quantity can greatly affect how easily they can be seen (Cao *et al.*, 2013).Third, the breakdown of pigment systems. Disease infections and pest attacks frequently result in the breakdown of chloroplasts or other organelles, which can alter the amounts of pigments such as anthocyanins, carotenoid pigments, and chlorophyll. In order to identify this kind of response, hyperspectral RS data is typically necessary (Grisham *et al.*, 2010). It should be recognized that, the presence of plant diseases or pests can result in one or more types of impairment. Various forms of symptoms may overlap or interact with one another and manifest at varying severities throughout different stages of the disease infection or pest attack because these processes often behave as temporal processes (Shi *et al.*, 2018). As a result, it's critical to recognize particular RS characteristics at various phases of the emergence of disease or pest infestation.

4.3 Detection of Plant Diseases and Insect Pests Using Artificial Intelligence

Currently, plant disease detection systems based on artificial intelligence (AI) are critical. They minimize crop diseases from developing frequently and the resulting losses. According to Jafar *et al.* (2024), the AI-based disease detection system adheres to preset procedures. In order to gather and record plant photos, the procedures entail a number of phases, one of which is placing different sensors in the agricultural area. After being gathered, the photos are divided and processed in order to be utilized as input in machine learning (ML) algorithms. Then the ML models forecast if a leaf is infected or healthy (Ayaz *et al.*, 2019). Figure 5.1 presents the framework of AI based detection system with preset processes to predict plant disease and insect pest.

4.3.1 Acquisition of Image

During this stage, pertinent photos of the diseased plant are taken and collected in order to use automated methods for classification. A picture is made up of binary data that can be edited and examined on a computer. High-resolution digital cameras are used to take pictures (Camargo and Smith, 2009). By capturing image samples in a variety of supported formats, including jpg, tif, and png, smartphones have shown to be helpful devices. They are then forwarded to the image preprocessing phase for adjustments before to usage. The picture acquisition stage is essential for precise disease and pest diagnosis. The literature on agricultural research has many well-known image files for different species of plants. It is feasible to investigate the impacts of various biotic stresses on plant health due to the datasets that comprise both healthy and diseased leaves. Online resources include IPM Images, Plant Doc, APS Images (Singh *et al.*, 2020), New Plant Diseases (Wani *et al.*, 2022), and many more datasets pertaining to plant infections.

4.3.2 Pre-processing of Image

Numerous elements, like noise, uneven or bright lighting, blur, undesirable backgrounds, etc., are present in the acquired photographs. As such, it is imperative to handle this unprocessed data and render it suitable for effective disease identification by automatic methods. After being transformed into a particular format, the raw data is cleaned up by eliminating distortion and noise. Images move on to the following stage, when the necessary processes of segmentation and feature extraction are completed. Standardization, color scaling, distortion reduction, picture size regularization, and noise removal are a few preprocessing techniques that enable scaling the image to the required dimensions at this point. The type, capacity, and value (HSV) approach is a common preprocessing technique in agricultural research that closely resembles the abilities of human observers (Jadhav and Patil, 2016). A common technique in agricultural research to enhance the clarity of image outline structures is the Laplacian filter. The Fourier transform (FT) filter effectively converts the pictures into the spatial frequency domain using a Fast Fourier Transform technique (Packa *et al.*, 2015). To diagnose an affected leaf with perfect precision, the image must be segmented.

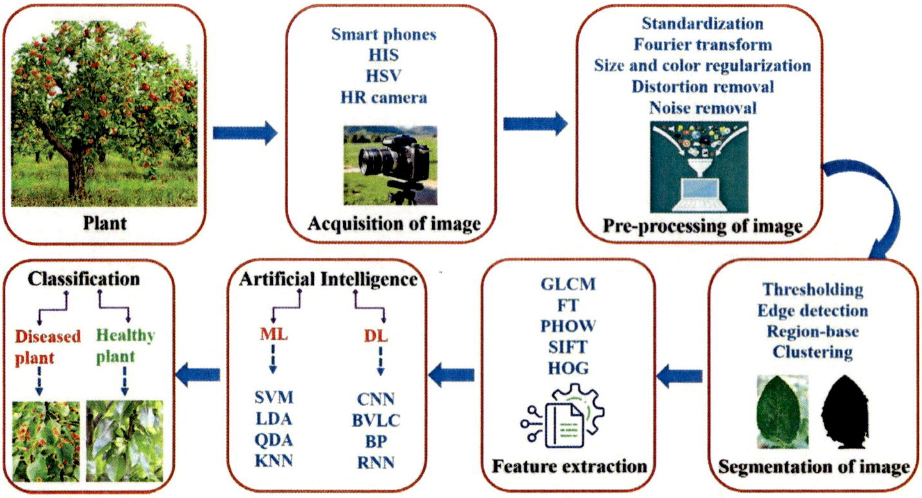

Fig. 4.1: Steps for detection of diseasesand pests in plants using AI

4.3.3 Segmentation of Image

A key method in agricultural research is segmentation, which involves breaking down an image into its constituent parts carefully to examine each one more closely and extract useful information that could improve our comprehension and knowledge (Jafar *et al.*, 2022). It is essential to segment the preprocessed pictures in order to classify unhealthy leaves and extract a variety of potentially useful attributes. To segment the provided images, conventional methods like thresholding, region-based, edge detection, and clustering rely on mathematical and image processing expertise. One of the best methods for image segmentation is thresholding, which divides images according to pixel intensity levels. Global, variable, and adaptive thresholding segmentation are its three varieties. Edge detection divides an image into segments according to its edges, which are usually referred to as the image's borders. The Canny edge detector, Sobel operator, and Laplacian of Gaussian (LoG) filter are a few well-known edge detection techniques. With region-based segmentation, the image is divided into several areas according to how similar the pixels are to each other in terms of form, color, and intensity value. Region Growing and Region Splitting are two popular region-based segmentation techniques (Aubry *et al.*, 2014). Pixels are grouped together through clustering based on color, shared texture, or other necessary characteristics. Two widely recognized clustering techniques for image segmentation include K-means (Ell and Sangwine, 2007) and Fuzzy C-means (Camargo and Smith, 2009).

4.3.4 Feature Extraction

Feature extraction is the process of obtaining features from unprocessed data (Basavaiah and Arlene, 2020). The attributes of color, shape, and texture are the feature descriptors of the input image. It is a crucial component of categorization tasks. The dataset is supplied as input to this step in the machine learning process in order to ascertain the health of the plants. Morphological features work better than others at identifying the damaged area on a leaf (Khirade and Patil, 2015). These properties can be obtained using a variety of techniques, including the color R moment, the color histogram, and others. Further features that have demonstrated increased efficacy include the Pyramid Histogram of Visual Words (PHOW), the Histogram of Oriented Gradients, and the Speed-up robust feature (Kaur *et al.*, 2019).

4.3.5 Detection and Classification of Plant Disease

In agricultural research, artificial intelligence (AI) is becoming progressively more significant, especially when it comes to diagnosing and categorizing plant diseases. The last step in this process is classification which involves grouping data into classes. This stage helps with detection and classification, particularly with identifying infected from healthy samples. Understanding the Deep Learning (DL) and Machine Learning (ML) classification and detection algorithms is necessary in order to be performed.

4.4 Machine Learning and Deep Learning Algorithms

Lately, plant pathology has used automated intelligence technology to detect infestations in plants. These technologies have the power to completely change how plant diseases are diagnosed and treated. Machine Learning (ML) is one of the most used AI technologies in plant pathology. Table 5.1 illustrates most frequently used algorithms used for detection of plant diseases and pests. Plant disease classification from digital photos has been done using machine learning (ML) methods such the c4.5 classifier, tree bagger, and linear support vector machines (Shoaib *et al.*, 2023). A large number of machine learning methods are crucial to modeling for pest and disease detection in plants. For instance, spectrum angle mapping (SAM) has been shown to be effective in monitoring wheat powdery mildew using a high-resolution SPOT-6 image (Yuan *et al.*, 2016) as well as in detecting the head blight disease in wheat using a hyperspectral imaging system (Bauriegel *et al.*, 2011). With the use of a Quickbird picture and maximum likelihood classification (MLC), a 91% mapping accuracy has been achieved for mountain pine beetle damage (Hicke and Logan, 2009).

Support vector machines (SVMs) are one of the machine learning algorithms that are worth mentioning as they have been used extensively in the monitoring of various pests and plant diseases. Romer *et al.* (2011) obtained 93% accuracy in early wheat leaf rust recognition using SVM classifiers and certain fluorescence characteristics. SVMs perform better than several other techniques, such as logistic regression, k-nearest neighbours (KNN), LDA, QDA, decision trees, and neural networks, in the detection of Huanglongbing (HLB) in citrus (Mishra *et al.*, 2011; Garcia-Ruiz *et al.*, 2013).

Table 4.1: Most common algorithms used for detection of plant diseases and insect pests

Type	Algorithm	Disease/ pest detected	Reference
Regression based models	Univariate linear regression (ULR)	Yellow rust of wheat	Huang *et al.*, 2007
	Multiple linear regression (MLR)	Rice leaf folder	Huang *et al.*, 2012
	Partial least squares regression (PLSR)	Bacterial spot of tomato	Jones *et al.*, 2010
Statistical discriminant analysis	Fisher linear discriminant analysis (FLDA)	Powdery mildew of wheat	Yuan *et al.*, 2014
	Quadratic discriminant analysis (QDA)	Septoria leaf blotch and Yellow rust of wheat	Moshou *et al.*, 2011
	Logistic regression	Rhizomania of sugar beet	Steddom *et al.*, 2003
Spectral unmixing algorithms	Linear spectral mixture analysis (LSMA)	Spider mite of cotton	Fitzgerald *et al.*, 2004
	Mixture tuned matched filtering (MTMF)	Yellow rust of wheat	Zhang *et al.*, 2014
Machine learning algorithms	Spectral angle mapping (SAM)	*Fusarium* head blight of wheat	Bauriegel *et al.*, 2011
	Random forest (RF)	Insect defoliators in Botswana	Adelabu *et al.*, 2014
	Support vector machines (SVM)	Wheat leaf rust	Romer *et al.*, 2011
	Artificial neural network (ANN)	Laurel wilt of avocado	De Castro *et al.*, 2015

Natural language processing, object recognition, and picture classification have all been profoundly impacted by deep learning (DL) technology (Sarker, 2021). Deep learning uses neural networks to choose features on its own, doing away with the need for labor-intensive artificial feature engineering. By integrating low-level data to create abstract, high-level features, it increases generalizability and accuracy in tasks like target identification and image recognition. Advanced algorithms have been developed, including autoencoders, deep belief networks (DBNs), convolutional neural networks (CNNs), etc.

Convolutional Architecture for Fast Feature Embedding or Caffe, is an open-source DL framework (BVLC) created by the Berkeley Vision and Learning Center and community members (Jia *et al.*, 2014). It is well-liked for classification tasks in images and videos, and is also thought to be a strong option because of its rapidity and effectiveness while training big models. Caffe's Python interface and C++ implementation make it simple to integrate with other Python libraries like SciPy and NumPy. This makes it possible to develop and experiment with DL models with a great deal of freedom. With an accuracy of 99.53%, CNNs identify and diagnose plant diseases based on a sizable quantity (n = 87,848) of leaf pictures (Ferentinos, 2018). It should be admitted, that one of the biggest obstacles to the widespread use of deep learning techniques is the vast quantity of labeled samples required. Therefore, in future research using deep learning approaches, gathering a large number of training examples concerning pests and plant diseases from RS data should be a top focus.

Computer vision (CV) is another AI technology that has been used in disease detection. Certain regions of interest in images, such as plant leaves and disease symptoms, can be recognized and localized using CV methods, such as object detection and semantic segmentation (Kurmi and Gangwar, 2022; Peng and Wang, 2022). These techniques can be combined with ML or DL algorithms for disease identification and classification, automatically converting the photos into recognized patterns or attributes. Nevertheless, CV algorithms may not be appropriate for diseases that have never been observed before and require a large amount of tagged image data for model training.

4.5 Challenges in Detecting Plant Diseases and Insect Pests

Despite the fact that the monitoring of plant diseases and pests has made positive strides in the previous few decades, there are still certain obstacles that make it difficult to put the procedures into reality. Future trends will be shaped by research on finding answers to these problems. The primary concern is the early identification of pests or plant diseases. Considering the AI, monitoring plant diseases and pests is typically completed once symptoms are completely manifested, which may be too late to direct preventative measures. It's critical to further explore the viability of fluorescence, SAR, thermal, and

Lidar RS observations and combine them with the sophisticated VIS-NIR RS in order to increase the early detection of infections (Li *et al.*, 2015). The second challenge is to precisely identify a particular pest or disease in practical field situations where multiple crop stresses could manifest at once. It is not always true to assume that every image depicts a single disease; in fact, several diseases, nutritional deficits, and pests might all coexist in one image at the same time. This complicates the identification and monitoring of a particular disease, because the way symptoms present varies depending on the specific region. As such, these elements must be taken into account while assessing photographs.

Furthermore, since the symptoms and patterns are similar, diagnosis can be difficult (Jafar *et al.*, 2024). Typically, visible spectrum research is conducted by researchers. While adding infrared spectral bands could aid in disease differentiation, there are additional costs and difficulties involved. Maintaining a close eye on the dynamics of diseases and insect pests at an incredibly fine resolution is the third problem. The AI systems must have a high enough resolution in all three dimensions viz., spatial, spectral, and temporal to accomplish this. At the moment, hardly a single solution can meet these needs. In addition, inclement weather presents a significant challenge to the ongoing gathering of optical pictures. Sharing of data and information is the fourth problem. Currently, a bottleneck in the modeling for plant disease and pest monitoring is the lack of adequate survey data. Thus, it is advised to start gathering data while the plants are being grown. For instance, using a smartphone app, farmers should be enlisted to log the prevalence and seriousness of pests and diseases in the fields. The pooled data must then be readily available in order to provide effective data mining and advanced algorithm-based model training.

4.6 Conclusion

Plant protection evaluation and management greatly depend on the timely, accurate, and effective monitoring of diseases and pests over large areas. Different artificial intelligence (AI) tools have been established in the past few decades to monitor pests and plant diseases. These systems have shown considerable potential to supplement manual inspection with less labor-intensive methods. CNNs have demonstrated outstanding performance in DL, while ML algorithms like SVM and RF have demonstrated impressive effectiveness in disease classification and detection. Complex disease and pest identification is now feasible due to image recognition advances. But the majority of the study in this field is restricted to laboratory experiments and mostly depends on gathered images of pests and plant diseases. Early detection of plant diseases and pests is essential forcontrolling and regulating their growth and spread. For effective detection and diagnosis, plant health and meteorological data must be incorporated. To fully utilize this technology, experts in plant protection and agriculture must work together, fusing their expertise with DL models and algorithms and incorporating the output into farming machinery. These methods, however, may not be appropriate for diseases that have not yet been identified and require a significant amount of annotated training data. In order to create generalizable models that can be used for many plant species and diseases, additional research is required. Likewise, more information must be made publicly available so that the models can be trained and assessed.

References

Adelabu, S., Mutanga, O., & Adam, E. (2014). Evaluating the impact of red-edge band from Rapideye image for classifying insect defoliation levels. ISPRS Journal of Photogrammetry and Remote Sensing, 95, 34-41.

Angermueller, C., Parnamaa, T., Parts, L., &Stegle, O. (2016). Deep learning for computational biology. Molecular Systems Biology, 12(7), 878.

Aubry, M., Paris, S., Hasinoff, S. W., Kautz, J., & Durand, F. (2014). Fast local laplacian filters: Theory and applications. ACM Transactions on Graphics (TOG), 33(5), 1-14.

Ayaz, M., Ammad-Uddin, M., Sharif, Z., Mansour, A., &Aggoune, E. H. M. (2019). Internet-of-things (IoT)-based smart agriculture: toward making the fields talk. IEEE Access, 7, 129551–129583.

Basavaiah, J., & Arlene Anthony, A. (2020). Tomato leaf disease classification using multiple feature extraction techniques. Wireless Personal Communications, 115(1), 633-651.

Bauriegel, E., Giebel, A., Geyer, M., Schmidt, U., &Herppich, W. B. (2011). Early detection of Fusariuminfection in wheat using hyper-spectral imaging. Computers and Electronics in Agriculture, 75(2), 304-312.

Camargo, A., & Smith, J. S. (2009). An image-processing based algorithm to automatically identify plant disease visual symptoms. Biosystems Engineering, 102(1), 9-21.

Cao, X., Luo, Y., Zhou, Y., Duan, X., & Cheng, D. (2013). Detection of powdery mildew in two winter wheat cultivars using canopy hyperspectral reflectance. Crop Protection, 45, 124-131.

De Castro, A. I., Ehsani, R., Ploetz, R., Crane, J. H., &Abdulridha, J. (2015). Optimum spectral and geometric parameters for early detection of laurel wilt disease in avocado. Remote Sensing of Environment, 171, 33-44.

Eli-Chukwu, N. C. (2019). Applications of artificial intelligence in agriculture: A review. Engineering, Technology & Applied Science Research, 9(4).

Ell, T. A., &Sangwine, S. J. (2007). Hypercomplex Fourier transforms of color images. IEEE Transactions on Image Processing, 16(1), 22-35.

Eraslan, G., Avsec, Z., Gagneur, J., &Theis, F. J. (2019). Deep learning: new computational modelling techniques for genomics. Nature Reviews Genetics, 20(7), 389-403.

Ferentinos, K. P. (2018). Deep learning models for plant disease detection and diagnosis. Computers and Electronics in Agriculture, 145, 311-318.

Fitzgerald, G. J., Maas, S. J., &Detar, W. R. (2004). Spider mite detection and canopy component mapping in cotton using hyperspectral imagery and spectral mixture analysis. Precision Agriculture, 5, 275-289.

Garcia-Ruiz, F., Sankaran, S., Maja, J. M., Lee, W. S., Rasmussen, J., &Ehsani, R. (2013). Comparison of two aerial imaging platforms for identification of Huanglongbing-infected citrus trees. Computers and Electronics in Agriculture, 91, 106-115.

Grisham, M. P., Johnson, R. M., &Zimba, P. V. (2010). Detecting Sugarcane yellow leaf virus infection in asymptomatic leaves with hyperspectral remote sensing and associated leaf pigment changes. Journal of Virological Methods, 167(2), 140-145.

Hicke, J. A., & Logan, J. (2009). Mapping whitebark pine mortality caused by a mountain pine beetle outbreak with high spatial resolution satellite imagery. International Journal of Remote Sensing, 30(17), 4427-4441.

Huang, J., Liao, H., Zhu, Y., Sun, J., Sun, Q., & Liu, X. (2012). Hyperspectral detection of rice damaged by rice leaf folder (Cnaphalocrocismedinalis). Computers and Electronics in Agriculture, 82, 100-107.

Huang, W., Lamb, D. W., Niu, Z., Zhang, Y., Liu, L., & Wang, J. (2007). Identification of yellow rust in wheat using in-situ spectral reflectance measurements and airborne hyperspectral imaging. Precision Agriculture, 8, 187-197.

Jadhav, S. B., &Patil, S. B. (2016). Grading of soybean leaf disease based on segmented image using k-means clustering. International Journal of Artificial Intelligence, 5, 13–215.

Jafar, A., Bibi, N., Naqvi, R. A., Sadeghi-Niaraki, A., &Jeong, D. (2024). Revolutionizing agriculture with artificial intelligence: plant disease detection methods, applications, and their limitations. Frontiers in Plant Science, 15, 1356260.

Jafar, A., Hameed, M. T., Akram, N., Waqas, U., Kim, H. S., & Naqvi, R. A. (2022). CardioNet: Automatic semantic segmentation to calculate the cardiothoracic ratio for cardiomegaly and other chest diseases. Journal of Personalized Medicine, 12(6), 988.

Jia, Y., Shelhamer, E., Donahue, J., Karayev, S., Long, J., Girshick, R., Guadarrama, S., & Darrell, T. (2014). Caffe: Convolutional architecture for fast feature embedding. In Proceedings of the 22nd ACM International Conference on Multimedia (pp. 675-678).

Jones, C. D., Jones, J. B., & Lee, W. S. (2010). Diagnosis of bacterial spot of tomato using spectral signatures. Computers and Electronics in Agriculture, 74(2), 329-335.

Kaur, S., Pandey, S., &Goel, S. (2019). Plants disease identification and classification through leaf images: A survey. Archives of Computational Methods in Engineering, 26, 507-530.

Khandelwal, P. M., &Chavhan, H. (2019). Artificial intelligence in agriculture: An emerging era of research. Research Gate Publication, 1(1), 56-60.

Khirade, S. D., &Patil, A. B. (2015). Plant disease detection using image processing. In 2015 International Conference on Computing Communication Control and Automation, pp. 768-771.

Kumar, A., Sarkar, S., & Pradhan, C. (2019). Recommendation system for crop identification and pest control technique in agriculture. In 2019 International Conference on Communication and Signal Processing (ICCSP) (pp. 0185-0189). IEEE. doi: 10.1109/ICCSP.2019.8698099.

Kumar, S., & Singh, S. K. (2020). Occluded thermal face recognition using bag of CNN ($ Bo $ CNN). IEEE Signal Processing Letters, 27, 975-979.

Kumar, V., &Vani, K. (2018). Agricultural robot: leaf disease detection and monitoring the field condition using machine learning and image processing. International Journal of Computational Intelligence Research, 14(7), 551-561.

Kurmi, Y., &Gangwar, S. (2022). A leaf image localization based algorithm for different crops disease classification. Information Processing in Agriculture, 9(3), 456-474.

Li, H., Zhao, C., Yang, G., & Feng, H. (2015). Variations in crop variables within wheat canopies and responses of canopy spectral characteristics and derived vegetation indices to different vertical leaf layers and spikes. Remote Sensing of Environment, 169, 358-374.

Li, J., Mi, Y., Li, G., &Ju, Z. (2019). CNN-based facial expression recognition from annotated rgb-d images for human–robot interaction. International Journal of Humanoid Robotics, 16(04), 1941002.

Liu, J., & Wang, X. (2021). Plant diseases and pests detection based on deep learning: a review. Plant Methods, 17, 1-18.

Mahlein, A. K. (2016). Plant disease detection by imaging sensors–parallels and specific demands for precision agriculture and plant phenotyping. Plant Disease, 100(2), 241-251.

Melnyk, P., You, Z., & Li, K. (2020). A high-performance CNN method for offline handwritten Chinese character recognition and visualization. Soft Computing, 24(11), 7977-7987.

Mishra, A., Karimi, D., Ehsani, R., &Albrigo, L. G. (2011). Evaluation of an active optical sensor for detection of Huanglongbing (HLB) disease. Biosystems Engineering, 110(3), 302-309.

Moshou, D., Bravo, C., Oberti, R., West, J. S., Ramon, H., Vougioukas, S., &Bochtis, D. (2011). Intelligent multi-sensor system for the detection and treatment of fungal diseases in arable crops. Biosystems Engineering, 108(4), 311-321.

Murugesan, R., Sudarsanam, S. K., Malathi, G., Vijayakumar, V., Neelanarayanan, V., Venugopal, R., Rekha, D., Saha, S., Bajaj, R., Miral, A., &Malolan, V. (2019). Artificial intelligence and agriculture 5. 0. International Journal of Recent Technology and Engineering, 8, 1870-1877.

Oerke, E. C. (2006). Crop losses to pests. The Journal of Agricultural Science, 144(1), 31-43.

Packa, D., Wiwart, M., Suchowilska, E., &Bienkowska, T. (2015). Morpho-anatomical traits of two lowest internodes related to lodging resistance in selected genotypes of Triticum. International Agrophysics, 29(4).

Peng, Y., & Wang, Y. (2022). Leaf disease image retrieval with object detection and deep metric learning. Frontiers in Plant Science, 13, 963302.

Romer, C., Burling, K., Hunsche, M., Rumpf, T., Noga, G., &Plumer, L. (2011). Robust fitting of fluorescence spectra for pre-symptomatic wheat leaf rust detection with support vector machines. Computers and Electronics in Agriculture, 79(2), 180-188.

Sarker, I. H. (2021). Deep learning: a comprehensive overview on techniques, taxonomy, applications and research directions. SN Computer Science, 2(6), 420.

Savary, S., Willocquet, L., Pethybridge, S. J., Esker, P., McRoberts, N., & Nelson, A. (2019). The global burden of pathogens and pests on major food crops. Nature Ecology & Evolution, 3(3), 430-439.

Shi, Y., Huang, W., Gonzalez-Moreno, P., Luke, B., Dong, Y., Zheng, Q., Ma, H., & Liu, L. (2018). Wavelet-based rust spectral feature set (WRSFs): A novel spectral feature set based on continuous wavelet transformation for tracking progressive host–pathogen interaction of yellow rust on wheat. Remote Sensing, 10(4), 525.

Shoaib, M., Shah, B., Ei-Sappagh, S., Ali, A., Ullah, A., Alenezi, F., Gechev, T., Hussain, T., & Ali, F. (2023). An advanced deep learning models-based plant disease detection: A review of recent research. Frontiers in Plant Science, 14, 1158933.

Singh, D., Jain, N., Jain, P., Kayal, P., Kumawat, S., &Batra, N. (2020). PlantDoc: A dataset for visual plant disease detection. In Proceedings of the 7th ACM IKDD CoDS and 25th COMAD (pp. 249-253).

Steddom, K., Heidel, G., Jones, D., & Rush, C. M. (2003). Remote detection of rhizomania in sugar beets. Phytopathology, 93(6), 720-726.

Strange, R. N., & Scott, P. R. (2005). Plant disease: a threat to global food security. Annual Review of Phytopathology, 43, 83-116.

Sujawat, G. S. (2021). Application of Artificial Intelligence in detection of diseases in plants: A Survey. Turkish Journal of Computer and Mathematics Education, 12(3), 3301-3305.

Sundararajan, S. K., Sankaragomathi, B., &Priya, D. S. (2019). Deep belief CNN feature representation based content based image retrieval for medical images. Journal of Medical Systems, 43(6), 174.

Talaviya, T., Shah, D., Patel, N., Yagnik, H., & Shah, M. (2020). Implementation of artificial intelligence in agriculture for optimisation of irrigation and application of pesticides and herbicides. Artificial Intelligence in Agriculture, 4, 58-73.

Toscano-Miranda, R., Toro, M., Aguilar, J., Caro, M., Marulanda, A., &Trebilcok, A. (2022). Artificial-intelligence and sensing techniques for the management of insect pests and diseases in cotton: A systematic literature review. The Journal of Agricultural Science, 160(1-2), 16-31.

Tsaftaris, S. A., Minervini, M., &Scharr, H. (2016). Machine learning for plant phenotyping needs image processing. Trends in Plant Science, 21(12), 989-991.

Wani, J. A., Sharma, S., Muzamil, M., Ahmed, S., Sharma, S., & Singh, S. (2022). Machine learning and deep learning based computational techniques in automatic agricultural diseases detection: Methodologies, applications, and challenges. Archives of Computational Methods in Engineering, 29(1), 641-677.

Yang, D., Li, S., Peng, Z., Wang, P., Wang, J., & Yang, H. (2019). MF-CNN: traffic flow prediction using convolutional neural network and multi-features fusion. IEICE Transactions on Information and Systems, 102(8), 1526-1536.

Yuan, L., Pu, R., Zhang, J., Wang, J., & Yang, H. (2016). Using high spatial resolution satellite imagery for mapping powdery mildew at a regional scale. Precision Agriculture, 17, 332-348.

Yuan, L., Zhang, J., Shi, Y., Nie, C., Wei, L., & Wang, J. (2014). Damage mapping of powdery mildew in winter wheat with high-resolution satellite image. Remote Sensing, 6(5), 3611-3623.

Zhang, J., Huang, Y., Yuan, L., Yang, G., Chen, L., & Zhao, C. (2015). Using satellite multi-spectral imagery for damage mapping of armyworm (Spodopterafrugiperda) in maize at a regional scale. Pest Management Science, 72, 335–348.

Zhang, J., Pu, R., Loraamm, R. W., Yang, G., & Wang, J. (2014). Comparison between wavelet spectral features and conventional spectral features in detecting yellow rust for winter wheat. Computers and Electronics in Agriculture, 100, 79-87.

5

Intelligent Spraying in Horticultural Crops

Kishor Varotariya[1] and Neha Sharma[2] and Ume Kulsum[3]

*[1]Division of Vegetable Science, Indian Institute of Horticulture
Bengaluru, 560089, India*

*[2]Division of Vegetable Science, Indian Agricultural Research Institute
New Delhi-110012, India*

[3]Division of Agricultural Sciences, Sher-e Kashmir University of Agricultural Sciences and Technology, Shalimar, Srinagar-190025, J&K, India

Abstract

Intelligent spraying systems in horticulture aim to enhance chemical efficacy, reduce costs and labour hazards, and minimize environmental damage. Variable-rate spray applications, utilizing advanced intelligent control systems, can significantly cut pesticide use and off-target pollution. Real-time variable-rate spraying technology ensures effective pesticide application, targeting only the necessary areas based on canopy size, season, and plant growth phase. Recent advancements in target detection systems, including laser, vision scanning, ultrasound, infrared, and spectrum systems, have improved precision in spraying. Machine vision and laser scanners are particularly effective in generating three-dimensional images and maps of plants and canopies. This chapters various sensing technologies for determining canopy structural parameters and their application in precision spraying, addressing the challenges and considerations in their use.

Keywords: *Horticulture, Intelligent spraying, Variable-rate application, Target detection systems, Precision spraying, Sensing technologies*

5.1 Introduction

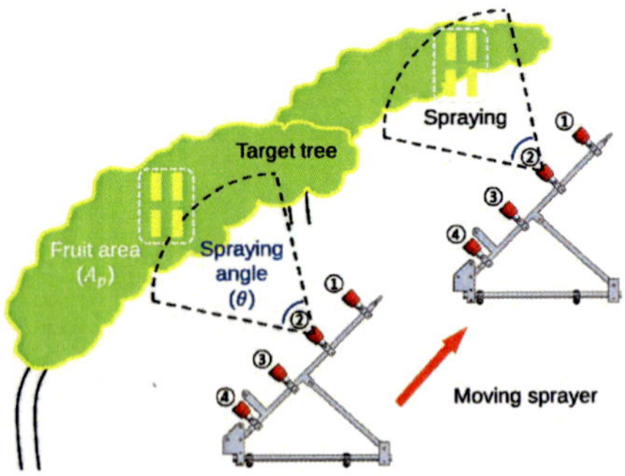

The application of chemical pesticides is crucial in horticultural production, but it is also one of the most hazardous agricultural operations. Overapplication can lead to significant environmental concerns. Spraying is vital for reducing harvest losses and improving productivity. According to Cho *et al.,*(2012), 30-35% of production damages can be prevented by reducing harmful insects and diseases through spraying. Plant protection is essential for ensuring yield quantity and quality. Effective pesticide use is crucial, as excessive application can lead to pests developing immunity, destruction of beneficial insects and natural pest enemies, and increased harmful insect populations. Excessive pesticide use also disrupts pollination, preventing fruit formation. The method of pesticide application is key to targeting the correct quantity and quality, reducing environmental contamination, and increasing sustainability (Abbas *et al.,* 2020). Site-specific application rates in precision horticulture can decrease pesticide use. Only 30-40% of pesticides typically reach their targets, with most lost to the environment, contaminating workers and causing pollution. Protecting plants with chemicals aims to prevent infections and ensure healthy growth.

Spray drift can harm non-target plants and contaminate watercourses, posing a severe and costly problem for horticultural farmers using air-assisted variable sprayers (Seol *et al.,* 2022). Spray drift refers to airborne spray droplets moving beyond the intended application area from aerial or ground-based methods. Studies have found that a significant portion of pesticides is lost to the air and ground, with up to 30-50% of the applied pesticides being lost through various means (Abbas *et al.,* 2020). Modern trends emphasize the importance of sensors and precision agriculture for effective pesticide use. Understanding tree canopy physical features is crucial for effective spray application. This study focuses on recognizing and geometrically classifying canopies for adequate pesticide protection. Variable rate spraying targets only the canopy, enhancing pesticide application efficiency (Rathnayake *et al.,* 2022). Electronic methods and innovative equipment reduce operational and environmental costs. Sensors tracking canopy characteristics and spacing ensure optimal use of inputs, promoting sustainability. The TRV (tree row volume) model describes the geometric structure of plant canopies, aiding in the adjustment of spraying methods to real-time canopy density and improving pesticide deposition. Sensor-based variable rate systems have been developed for crops like cabbage, spraying only in the presence of foliage (Chen *et al.*, 2020). Growing interest in spray technology aims to reduce contamination during application. Real-time data controls pesticide application at variable flow rates, increasing efficiency and reducing environmental pollution. Current technology minimizes pesticide use and losses, benefiting farmers, consumers, and the environment. Sensors measure target parameters, and calculated pesticide amounts are applied through nozzles using data processing units.

The development of the intelligent sprayer prototype for nurseries and orchards represents a significant advancement in horticultural practices. This system offers numerous advantages compared to conventional chemical application methods. One key feature is its utilization of variable-rate control technologies, which enable the adjustment of spray output according to the specific needs of target canopies (Deshmukh *et al.*, 2021). By tailoring the amount of spray applied to individual plants or areas based on factors such as canopy density and growth stage, the intelligent sprayer maximizes efficiency while minimizing waste.

Table 5.1. Ideal test plants, and pests of study for conventional constant-rate and intelligent variable-rate spray applications

Plant	Conventional Spray		Intelligent Spray		Pests of Study
	Rows (No.)	Plot size (acress)	Rows (No.)	Plot size (acress)	
Apple	10	1.90	8	1.78	Codling moth, Oriental fruit moth, scab, powdery mildew
Peach	5	1.41	6	1.41	Oriental fruit moth, brown rot, powdery mildew
Blueberry	5	0.72	6	0.86	Spotted wing drosophila mummy berry, phonopsis
Black raspberry	7	1.41	8	1.33	Anthracnose

One of the primary benefits of this approach is the reduction in chemical costs for growers. By optimizing spray application and minimizing overuse, growers can achieve substantial savings in chemical inputs. Additionally, the intelligent sprayer significantly mitigates the risk of environmental contamination. With fewer chemicals missing the target and dispersing into the surrounding environment, the potential for contamination of water sources, soil, and non-target organisms is greatly reduced (Kim *et al.*, 2020). This not only promotes environmental sustainability but also ensures compliance with regulatory standards and safeguards the health of ecosystems. Overall, the intelligent sprayer prototype represents a promising solution for improving pesticide application practices in horticulture. Its innovative features offer both economic and environmental benefits, making it a valuable tool for modern agricultural production.

5.2 Basic Principle of Variable Rate Spray Technology in Horticulture

Variable rate spray technology in horticulture leverages sensors to detect the target plant and its geometrical structural features, collecting and processing data for precise application (Chen *et al.,* 2019). These sensors, which may include optical, ultrasound, or infrared types, measure the dimensions and density of the plant canopy. The gathered data is transmitted to a control unit that adjusts the spraying rate of the device in real-time.

The control unit's real-time adjustments ensure that the amount of pesticide or nutrient sprayed corresponds precisely to the plant's structural requirements. This adaptability is crucial as the canopy and leaf density vary across different parts of the plant, necessitating targeted spraying to maximize efficacy and minimize waste.

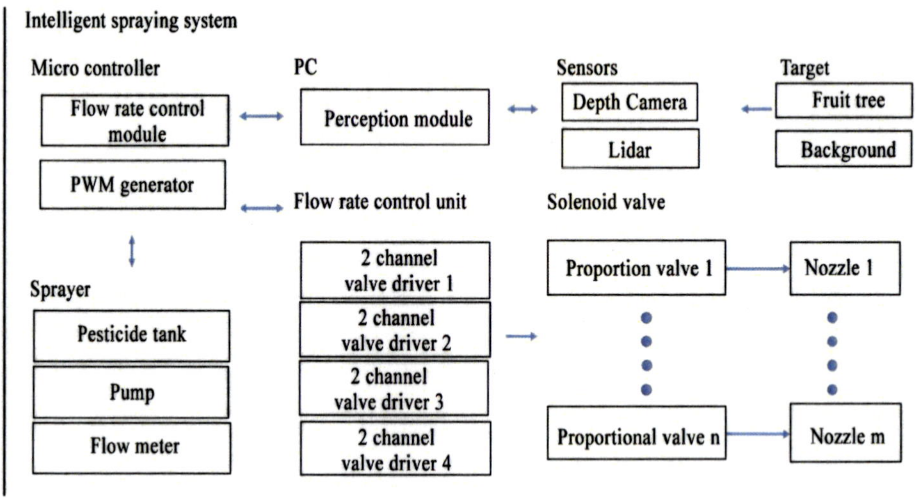

Fig. 5.1 Field channel of a Intelligent Spraying Robot with Flow Control for Horticultural Orchards (Hejazipoor *et al.*, 2021).

By accurately matching the spray amount to the plant's geometric structure, variable rate spray technology enhances the efficiency of pesticide and nutrient applications. It ensures that each part of the canopy receives the appropriate amount of spray, thereby improving pest and disease control, enhancing plant health, and promoting sustainable agricultural practices. This technology reduces chemical usage, minimizes environmental impact, and lowers costs, providing significant benefits for horticultural production (Vathsala *et al.,*2023).

5.3 Key Components of intelligent Spraying Systems in Horticulture

Smart spraying technology in horticulture integrates various advanced hardware and software capabilities to optimize agricultural chemical applications (Kumar *et al.,* 2023):

- Sensors: These measure real-time data such as air temperature, wind speed, crop canopy height, leaf wetness, and soil moisture. Optical sensors identify weed patches or pest infestations, triggering targeted spraying. Key sensor types include ultrasonic, optical, capacitive, photoelectric, and hybrid sensors.
- Control Systems: These systems respond to sensor data by modulating nozzle flow rates, droplet sizes, sprayer boom heights, and positioning, enabling variable rate spraying tailored to field variability.
- Variable Rate Technology: This technology adjusts application rates across different zones in a field based on specific requirements determined through sensors, GPS mapping, and yield data analysis. Major components include flow control nozzles, control valves, and variable rate pumps.
- GPS and Mapping: GPS locates the sprayer's position, while GIS generates application maps showing areas that have been sprayed, helping to avoid redundancy.
- Robotics: Unmanned autonomous vehicles equipped with sensors scan fields and conduct targeted spraying without operators onboard, enhancing safety and efficiency.
- Connectivity: Wireless networking and Internet of Things (IoT) connectivity enable remote monitoring of spraying operations through mobile apps and web dashboards, supporting telematics and real-time analytics.

5.4 Characteristics of Intelligent Spray Technology in Horticulture

Intelligent spray technology in horticulture is defined by its precision, efficiency, and sustainability. Utilizing advanced sensors and real-time data processing, this technology enhances the application of pesticides and nutrients (Hejazipoor *et al.,* 2021).

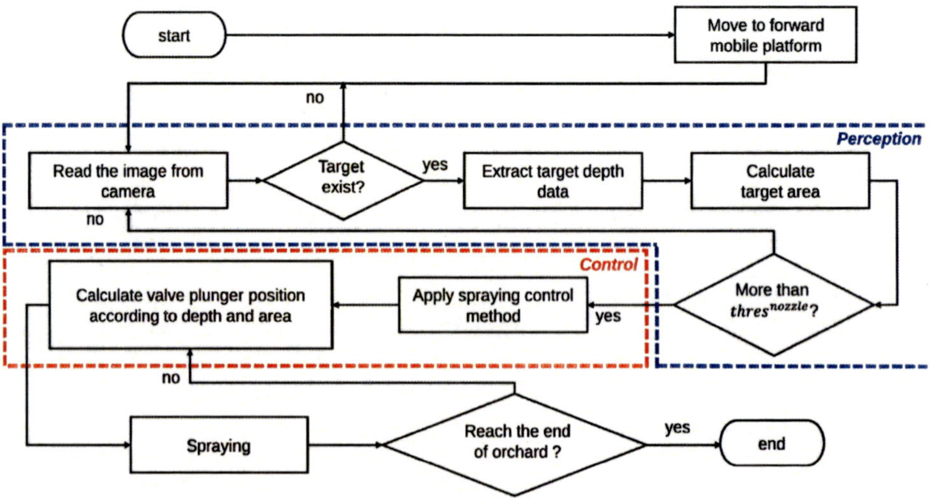

Fig. 5.2 Flow chart of intelligent spraying system

1. Precision Targeting: Intelligent spray systems employ sensors such as optical, ultrasound, and infrared to accurately measure plant canopy dimensions and leaf density. This precise targeting ensures that the right amount of spray is applied only where needed, reducing waste and increasing efficacy.
2. Real-Time Adjustments: The control units in intelligent spray systems process sensor data on-the-fly, adjusting the spray output in real-time based on the plant's structural features. This adaptability ensures consistent and optimized coverage across varied canopy structures.
3. Environmental Sustainability: By reducing over-application and ensuring that chemicals are applied only to the target areas, intelligent spray technology minimizes environmental impact. This precision reduces chemical runoff, protecting water sources and non-target organisms.
4. Cost Efficiency: Reduced chemical usage translates to lower input costs for farmers. Moreover, improved pest and disease control can lead to higher yields and better quality produce, offering economic advantages.
5. Adaptability and Versatility: Intelligent spray systems can be customized for different crops and canopy structures, making them suitable for a wide range of horticultural applications, from fruit orchards to vegetable fields.

Overall, intelligent spray technology represents a major advancement in horticulture, promoting sustainable farming practices while enhancing the precision and effectiveness of crop management.

5.5 Real-Time Sensor Systems for Plant Canopy Detection

The design of real-time sensor systems to detect the target and geometrical structure of plant canopies has significantly advanced. These sensor systems convert measured physical dimensions into analogue electrical data, which is then processed for various applications in precision agriculture (Cho *et al.,* 2012). The operation of sensor systems relies on interaction with the environment, where the feedback from this interaction is converted into an output signal, organized through a technical procedure.

Several physical phenomena can be harnessed in sensor production. For instance, optical sensors utilize light reflection and absorption properties to gauge canopy structure and health. These sensors often employ laser or infrared technology to measure the distance and density of plant foliage, providing three-dimensional maps of the canopy (Deshmukh *et al.,* 2021). Ultrasound sensors, on the other hand, use sound waves to detect the spatial characteristics of plants. They are particularly useful for measuring the height and volume of plant canopies.

Infrared and spectral sensors analyse the light spectrum reflected from plant surfaces to assess their health and detect stress indicators, such as water deficiency or pest infestation. These sensors are integral in implementing site-specific management practices, enabling precise application of inputs like water, fertilizers, and pesticides (Song *et al.,* 2020). Advanced sensor systems are often integrated with machine learning algorithms, enhancing their ability to interpret complex data patterns and make real-time decisions. For example, deep learning models can analyse sensor data to optimize spraying patterns in orchards, ensuring efficient pesticide use and minimizing environmental impact (Hejazipoor *et al.,* 2021).

5.6 Aids of Intelligent Spraying

5.6.1 Infrared Sensor in Intelligent Spraying

In horticulture, infrared sensor systems play a crucial role in various applications, offering automated object recognition capabilities and aiding in precision agriculture practices. These sensors detect the presence of objects based on the infrared radiation emitted by all objects with temperatures above absolute zero (Song *et al.,* 2020). By transmitting and receiving high-intensity light pulses, infrared sensors effectively detect objects within their range. In the context of horticulture, these sensors are utilized for diverse purposes, including monitoring plant health, detecting pests and diseases, and controlling automated equipment such as irrigation systems and harvesting robots.

Measuring distances using infrared detection often requires advanced techniques like phase transition measurement or triangulation, ensuring accurate and reliable results. This precision is essential in horticultural settings, where precise measurements are crucial for optimizing resource allocation and maximizing crop yields (Bzhikhatloy and Pshenokova, 2023). Moreover, infrared sensor systems enable automated object recognition, facilitating the identification of specific targets such as weeds, diseased plants, or ripe fruits. This capability enhances efficiency and productivity in horticultural operations by enabling timely interventions and targeted treatments, ultimately contributing to improved crop quality and yield. Overall, infrared sensor technology is indispensable in modern horticulture, offering valuable insights and automation capabilities to enhance agricultural practices and outcomes.

5.6.2 Ultrasonic Sensor in Intelligent Spraying

Ultrasonic sensor systems provide rapid and automated distance measurements in horticulture. These sensors measure time delay information to calculate distances, with the delay between transmission and sound reflection proportional to the distance from the sensor to the object. Utilizing the speed of sound, ultrasonic sensors accurately determine distances. Comprising a transmitter and receiver, these sensors emit ultrasonic pulses and capture their reflections. In horticulture, ultrasonic sensor systems are valuable for various applications, including monitoring plant growth, detecting obstacles in automated equipment, and optimizing irrigation systems (Geng *et al.,* 2012). Their rapid and precise measurements contribute to efficient and effective crop management practices, ultimately enhancing productivity and yield.

5.6.3 LIDAR Sensor in Intelligent Spraying

LIDAR (Light Detection and Ranging) sensor systems play a pivotal role in horticulture by providing remote sensing capabilities. These systems utilize laser pulses to collect precise measurements, enabling the creation of detailed 3D models and maps of objects and environments (Hanzipoor*et al.,* 2021). Acting as active sensors, LIDAR emits modulated laser radiation towards targets, which then reflect the radiation back to an optical receiver. Distance is calculated based on the flight time of the pulsed light, measuring the time it takes for the emitted light to travel to the target and back. In horticulture, LIDAR technology aids in precision farming practices, including crop monitoring, yield estimation, and terrain mapping.

5.6.4 Machine Vision System in Intelligent Spraying

In horticulture, machine vision systems offer efficient recognition of vegetation characteristics such as position, size, shape, colour, and texture.

This technology leverages computer vision to interpret images acquired by one or more video cameras. By enabling computers to "see," machine vision systems support various applications in mechanized agriculture.

Type of spraying	Size of the droplets
Conventional spraying (through nozzles)	100–300 microns
Large-volume spraying	300 microns
Low-volume spraying	100–200 microns
Ultra-high volume spraying	100 microns
Cold mist	50–100 microns
Hot mist	1–20 microns
Dry mist	1–5 microns

They are particularly popular for robotic applications, aiding in tasks such as plant monitoring, fruit harvesting, and weed detection (Hanzipoor*et al.,* 2021). Machine vision enhances automation in horticulture by providing real-time insights into crop health, facilitating precise interventions, and optimizing resource utilization. Overall, machine vision is instrumental in improving efficiency and productivity in horticultural practices.

5.6.5 Sensor Fusion System in Intelligent Spraying

Sensor fusion systems integrate data from multiple sources to optimize information processing. Sensor fusion involves the intelligent integration of data from various sensors deployed in the field. These sensors provide diverse information, including plant health metrics, environmental conditions, and soil characteristics. By combining data from multiple sensors, sensor fusion systems enhance decision-making processes in horticulture. They enable growers to obtain a comprehensive understanding of crop conditions and environmental factors, facilitating more informed and effective management strategies (Bzhikhatloy and Pshenokova, 2023). Despite the challenges posed by noise and clutter in multi-sensor environments, sensor fusion systems offer valuable insights for improving crop yield, quality, and sustainability in horticultural practices.

5.6.6 Use Case Implementation of Sensor Technology

In this chapter, the implementation of sensor technology is exemplified through a machine vision spraying system, providing insight into its development process. The focus lies on crop bed detection within a greenhouse bed, a crucial function of the spraying system. This involves detecting plants and identifying spray point locations for targeted spraying.The hardware configuration of the automatic spraying platform system comprises three main components: the image vision unit, the control unit, and the spraying unit. The image vision unit captures images of the crop bed, while the control unit processes these images to identify plant locations and determine optimal spray points. The spraying unit then delivers targeted sprays based on the information provided by the control unit (Bzhikhatloy and Pshenokova, 2023).

Through this implementation, the chapter aims to illustrate the practical application of sensor technology in agricultural settings, specifically in improving efficiency and precision in pesticide application within greenhouse environments.

5.7 Smart Aerial Intelligent Spraying Systems

Traditionally, pesticide application in agriculture has involved the use of fixed-wing aircraft or helicopters with pilots on board. However, the advent of small, remotely piloted aircraft, particularly in Southeast Asia, is transforming this practice. In South Korea, drones account for about 30% of agricultural spraying, while in Japan, they are used for approximately 40% of rice crop spraying (Cho *et al.,* 2012). While drone usage for spraying is still emerging in the United States, interest in this technology is on the rise among pesticide applicators.

Table 5.2 Drone spray specification

Specification	**Models**			
	AG/D5L	**AG/D1OL**	**AG/D18L**	**AG/D18L**
FUSELAGE				
Number of Motors	6	6	6	6
Aircraft size (without propeller) (mm)	1100*1100 *550	1330*1330 *550	1900*1900 *650	1900x1900x650
Folded size of the aircraft, (mm)	700*700 *550	700*650 *550	700*700 *650	700x700x650
FCL Gross Weight/Net Weight (KG)	50/10.5	60/18	63/20.8	63/20.8
Maximum take-off weight of the aircraft (KG)	20	24	43	43
Economic load (KG)	5	10	20	25
Aircraft battery	10000m AH SS 25C			
Remote control battery	3000mAh 3.7V			
Motor	8318 120 KV	170 KV	8330 100 KV	170 KV
FLIGHT PARAMETERS				
Remote control frequency	2.4 GHz			
Control distance	1000M*			

Operating flight speed (m/s)	3-10 (adjustable)			
Maximum flight altitude	GPS mode 30M / attitude mode unlimited			
Hover accuracy	±0.8 M horizontal / ± 0.5 M vertical			
No-load flight time (Minutes)	18-23	18-23	15-20	18-23
Operating flight time (Minutes)	10-13	10-13	10-14	10-13
Flight mode	Manual mode/operation mode/GPS mode			
Flight operating temperature	0-50"			

Remotely piloted aircraft go by several names and acronyms, including (Yallappa *et al.,* 2017):

- Unmanned Aerial Vehicle (UAV)
- Unmanned Aerial System (UAS)
- Small Unmanned Aerial System (SUAS)
- Remotely Piloted Vehicle (RPV)
- Remotely Piloted Aircraft (RPA)
- Remotely Operated Aircraft (ROA)
- Remotely Piloted Aerial Application System (RPAAS)

Although UAV and UAS are widely used terms, "drone" is the most common term among the public and is used throughout this text(Yallappa*et al.,* 2017). This discussion covers the specifications of drone sprayers, their benefits for aerial spraying, and the challenges that impede their wider adoption by pesticide applicators.

Drones offer a solution to these problems by enabling precise and efficient crop monitoring and pesticide application, thereby enhancing crop yields without posing health hazards to humans. Agricultural drones, primarily autonomous, operate by sensing their environment and executing tasks assigned by human operators (Rathnayake *et al.,* 2022). These drones are equipped with storage tanks for liquid pesticides, connected to nozzles via pipes and electric pressure pumps, allowing for targeted spraying. The operation of these components can be controlled manually through a remote controller or automatically via pre-programmed commands to ensure precise pesticide application.Drones efficiently monitor plant growth by providing real-time data from the moment of plant emergence to harvest. Utilizing advanced GPS or GNSS technology, high-resolution cameras, and variable flying speeds and altitudes, drones offer comprehensive insights into crop and soil conditions.

The use of drones for pesticide spraying is particularly advantageous for several reasons:

1. Challenging topography or soil conditions that preclude traditional ground sprayers or conventional agricultural aircraft.
2. Lack of availability or prohibitive costs of airplanes and helicopters.
3. Greater efficiency in spraying small, irregularly shaped fields.
4. Improved precision and reduced environmental impact compared to traditional methods.

5.8 Benefits of Intelligentspraying Technology in Horticulture

Compared to conventional uniform spraying, smart precision spraying offers several key advantages (Zhu *et al.,* 2010):

1. Optimized Chemical Use: By targeting applications based on specific needs, smart spraying reduces chemical overuse, significantly cutting costs while maintaining effectiveness.
2. Higher Efficiency: Spot spraying for weeds and variable rate applications enhance efficiency by 20-40% compared to uniform spraying, ensuring more precise treatment.
3. Lower Operating Costs: Precision spraying minimizes redundant applications, reducing the amount of chemicals required per acre and lowering long-term costs substantially.
4. Reduced Environmental Impact: Targeted spraying minimizes off-target drift and runoff, preventing contamination of nearby water bodies by applying the correct amount of chemicals exactly where needed.
5. Increased Yields: Healthier crops result from early detection and treatment of pest infestations and diseases, along with reduced weed competition, leading to higher yields. 6. Operator Safety: Closed transfer systems reduce handler exposure to chemicals, and autonomous sprayers further enhance safety by limiting direct human interaction with pesticides.

5.9 Future Thrust

The future of intelligent spraying in horticulture is poised to revolutionize agricultural practices with a focus on sustainability, precision, and efficiency. Advances in sensor technologies, machine learning, and artificial intelligence will enable real-time monitoring and data analysis, ensuring precise pesticide application tailored to plant health and growth stages. Integrating drones and autonomous vehicles will further enhance the accuracy and coverage of spraying operations, reducing chemical usage and environmental impact.

Innovations in variable-rate technology will allow for the dynamic adjustment of spraying parameters based on real-time data from the field, optimizing the amount of pesticide used and minimizing waste. Furthermore, the adoption of Internet of Things (IoT) devices will facilitate seamless communication between different systems, enhancing the overall efficiency of pest and disease management.Overall, intelligent spraying will lead to healthier crops, higher yields, and reduced labor and chemical costs, fostering a more sustainable and productive horticulture industry.

5.10 Conclusion

Intelligent spraying systems in horticulture significantly enhance chemical application efficacy while reducing costs, labor hazards, and environmental damage. Utilizing advanced variable-rate and real-time control technologies, these systems precisely target necessary areas based on canopy size, season, and plant growth phase, cutting pesticide use and off-target pollution. Advancements in target detection, including laser, vision scanning, ultrasound, infrared, and spectrum systems, have markedly improved spraying precision. Machine vision and laser scanners, in particular, effectively generate three-dimensional images of plants and canopies. This chapter highlights the critical role of various sensing technologies in precision spraying and discusses the challenges and considerations involved in their implementation.

References

Abbas, I., Liu, J., Faheem, M., Noor, R. S., Shaikh, S. A., Solangi, K. A., & Raza, S. M. (2020). Different sensor based intelligent spraying systems in Agriculture. Sensors and Actuators A: Physical, 316, 112265.

Bzhikhatlov, K., & Pshenokova, I. (2023, June). Intelligent Spraying System of Autonomous Mobile Agricultural Robot. In International Conference on Agriculture Digitalization and Organic Production (pp. 269-278). Singapore: Springer Nature Singapore.

Chen, L., Wallhead, M., Reding, M., Horst, L., & Zhu, H. (2020). Control of insect pests and diseases in an Ohio fruit farm with a laser-guided intelligent sprayer. HortTechnology, 30(2), 168-175.

Chen, L., Wallhead, M., Zhu, H., & Fulcher, A. (2019). Control of insects and diseases with intelligent variable-rate sprayers in ornamental nurseries. Journal of Environmental Horticulture, 37(3), 90-100.

Cho, S., Takagi, K., Kwon, H., Seo, D., Ogawa, K., Kikuchi, K., & Kawasaki, A. (2012). Multi-walled carbon nanotube-reinforced copper nanocomposite coating fabricated by low-pressure cold spray process. Surface and coatings Technology, 206(16), 3488-3494.

Deshmukh, D., Pratihar, D. K., Deb, A. K., Ray, H., & Bhattacharyya, N. (2021). Design and development of intelligent pesticide spraying system for agricultural robot. In Hybrid Intelligent Systems: 20th International Conference on Hybrid Intelligent Systems (HIS 2020), December 14-16, 2020 (pp. 157-170). Springer International Publishing.

Geng, C., Zhang, K., Zhang, E., Zhang, J., & Li, W. (2012). Assessment on spraying effect of intelligent spraying robot by experiment. Transactions of the Chinese Society of Agricultural Engineering, 28(1), 114-117.

Hejazipoor, H., Massah, J., Soryani, M., Vakilian, K. A., & Chegini, G. (2021). An intelligent spraying robot based on plant bulk volume. Computers and Electronics in Agriculture, 180, 105859.

Kim, J., Seol, J., Lee, S., Hong, S. W., & Son, H. I. (2020, May). An intelligent spraying system with deep learning-based semantic segmentation of fruit trees in orchards. In 2020 IEEE international conference on robotics and automation (ICRA) (pp. 3923-3929). IEEE.

Kumar, V., Jakhwal, R., Chaudhary, N., & Singh, S. (2023). Artificial intelligence in horticulture crops. Annals of Horticulture, 16(1), 72-79.

Rathnayake, A. P., Sahni, R. K., Khot, L. R., Hoheisel, G. A., & Zhu, H. (2022). Intelligent sprayer spray rates optimization to efficiently apply chemicals in modern apple orchards. Journal of the ASABE, 65(6), 1411-1420.

Seol, J., Kim, J., & Son, H. I. (2022). Field evaluations of a deep learning-based intelligent spraying robot with flow control for pear orchards. Precision Agriculture, 23(2), 712-732.

Song, L., Huang, J., Liang, X., Yang, S. X., Hu, W., & Tang, D. (2020). An intelligent multi-sensor variable spray system with chaotic optimization and adaptive fuzzy control. Sensors, 20(10), 2954.

Vathsala, A. V., Lakshmi, H. N., Yedukondalu, G., Rao, C. R. S., Kotha, M., & Changala, R. (2023). Optimization of Irrigation and Herbicides Using Artificial Intelligence in Agriculture. International Journal of Intelligent Systems and Applications in Engineering, 11(3), 503-518.

Yallappa, D., Veerangouda, M., Maski, D., Palled, V., &Bheemanna, M. (2017, October). Development and evaluation of drone mounted sprayer for pesticide applications to crops. In 2017 IEEE global humanitarian technology conference (GHTC) (pp. 1-7). IEEE.

Zhu, H., Jeon, H. Y., Gu, J., Derksen, R. C., Krause, C. R., Ozkan, H. E., & Rosetta, R. (2010). Development of Two Intelligent Spray Systems for Ornamental Nurseries©. Proceedings of the International Plant Propagators' Society, Miami, FL, USA, 60, 322.

6

Robotic Pollination in Horticultural Crops

Suhail Quyoom Wani and Gazala Gulzar

Division of Plant Pathology, Sher-e-Kashmir University of Agricultural Sciences and Technology, Shalimar, Srinagar-190025, J&K, India

6.1 Introduction

Perennial crops are crucial for maintaining food security, despite making up only 4.2–4.7% of all agricultural acreage (Kreitzman *et al.*, 2020). The value of fruit and nut production reached a value of USD 823 billion globally in 2020, with the top five fruit crops being apples (87 Mt), oranges (78 Mt), grapes (77 Mt), mangoes (55 Mt), and apricots (41 Mt) (Statista, 2020). The tree nut business is growing quickly, although making up a smaller share of the world's agricultural output. Global tree nut output is expected to reach 4.2 Mt in 2019–2020, with almonds, cashews, walnuts, hazelnuts, and pistachios accounting for 31%, 17%, 21%, 12%, and 14% of the nut market share, respectively (Serna, 2020).

Tree crops require a longer investment period before producing a commercial yield; sweet cherry takes three years to attain a commercial yield, while pistachio takes ten (Rezaei *et al.*, 2019). High-density monocultures of cultivars chosen for their potential yield, fruit quality, and disease resistance are part of optimal tree crop production techniques. These monocultures are maintained under intensive, high-input farming systems (Simon *et al.*, 2017). However, inadequate pollination can seriously impair the quantity and quality of these agricultural products, prompting for meticulous dedication to pollination research and management (Bosch *et al.*, 2021; Forbes *et al.*, 2019).

Effective pollination, and on the contrary, situations in which pollination is restricted or deficient, depend on both internal and external factors (Figure 7.1, Table 7.1). The period of stigma receptivity, the lifespan of the ovule, and the quality and availability of compatible, suitable pollen are all considered as intrinsic factors (Howlett *et al.*, 2015). Agronomic methods pertaining to plant nutrition, orchard layout, pest and disease control, pollinator activity, pollen quality, and weather patterns and their effects on flowering synchrony are examples of extrinsic factors(Toledo-Hernandez *et al.*, 2017). Hatfield and Prueger assert that pollination is one of the phenological stages that is most sensitive to temperature, and as such, it has a significant impact on productivity (Hatfield and Prueger, 2015). Climate has been shown to be a key factor in determining pollination success and production in crops including peaches, apples, and almonds. Poor pollination regularly lowers crop quality and reduces production, impacting characteristics including symmetry of fruits, mineral content, and storability, as well as oil and vitamin E levels in almonds (Eyles *et al.*, 2022).

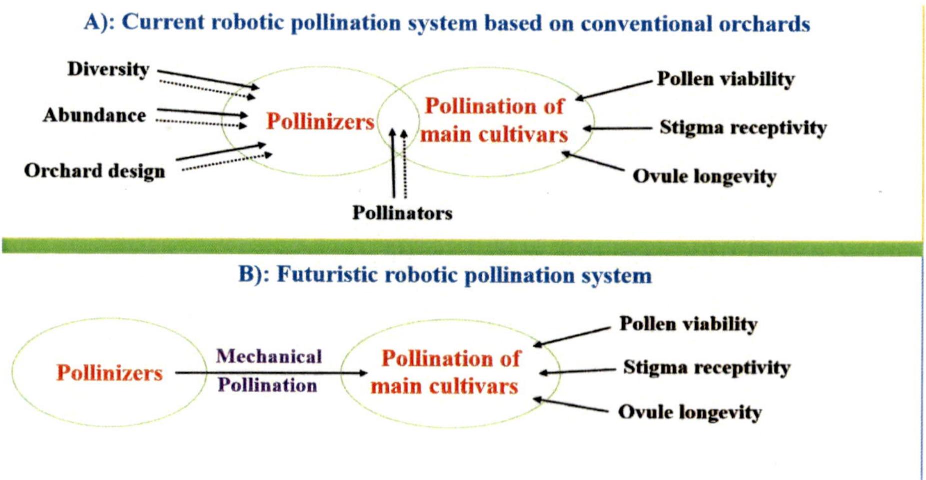

Fig. 6.1: Diagrammatic representation of the factors influencing cross-pollination of fruit and nut crops for a pollinator and pollinator model production system **A)**: with (solid line) or without (dashed line) a robotic pollination system. Variables with dotted lines would be less significant, or **B)**: Robotic pollination only when pollinizers are grown in a separate orchard. In both scenarios, the environmental elements and agronomic management strategies would be the same.

With the exception of some self-fertile cultivars and parthenocarpic crops like citrus, bananas, and common fig, most tree crops depend on insect assisted cross-pollination for successful pollination (FAO, 2018). In order to produce the best possible seed and fruit set, horticulturists use both managed and natural insect pollination (Isaacs *et al.*, 2017). As a result, a significant reliance on European honey bees (*Apismellifera*) from controlled hives or wild colonies has been established in order to increase the yield of commercial horticultural tree crops (Goodrich *et al.*, 2019). However, the production of fruits and nuts is seriously jeopardized by this dependence on a single pollinator species. A number of issues, such as the use of pesticides, disease outbreaks, migratory beekeeping, and habitat fragmentation, are contributing to the loss in both managed and wild bee populations (Traynor *et al.*, 2020). Additionally, a particularly catastrophic causative element has been discovered as the invasive parasite, mite *Varroa destructor*, which not only transmits bee viruses but also directly affects growing pupae and adult bees. Though research on *Varroa*has advanced (Traynor *et al.*, 2020), finding long-term control strategies still seems unlikely (Guichard *et al.*, 2020).

Table 6.1: The factors that contribute to lack of pollination in fruit crops and advantages of mechanical pollination (MP) systems

Factors leading to inadequate pollination	Advantages of mechanical pollination (MP) systems
Unfavorable weather can shorten flowering duration and intensity as well as decrease insect activity.	Safeguard against unfavorable weather circumstances because MP can be used day or night in undesirable weather conditions, but not in rainy ones (Whiting, 2019)
Asynchrony of flowering: Depending on environmental factors, the overlap of stigma receptivity and pollen changes from year to year (Cerovic *et al.*, 2020). Asynchrony in the flowering of suitable cultivars for monoecious species. Asynchrony between staminate and pistillate flowers in dioecious species.	Lesser requirement for pollinizers because pollination can happen when the main crop is available.
Inclination of honey bees towards substitute nectar and/or pollen sources over the primary crop. Reducing non-target food sources for pollinators, however, may result in lower nutrition and a decline in pollinator health (Quinlan *et al.*, 2021).	It is possible to assist both native and managed pollinators by encouraging diversity among non-target feed sources.
Certain crop species are unsuitable for honey bee pollination because their blooms do not appeal honey bees, either due to poor quality nectar or no nectar {e.g., pears (Quinet *et al.*, 2016); kiwifruit (Saez *et al.*, 2019)}.	Lessens reliance on native and managed pollinators.
A significant portion of sterile (Ascari *et al.*, 2022) or non-viable (Cerovic *et al.*, 2020) pollen can be produced by pollinizers. Pollen viability can also be lowered by unfavorable weather patterns and agronomic techniques (such as the use of pesticides) (Padilla *et al.*, 2017).	Pollen quality in storage can be tracked and obtained from orchards that yield viable and fertile pollen during a specific season.

Limitations in the pollination design, such as density (10–20%), dispersion (10–15 m range of compatible pollinizer) and ratio of male to female trees and in the right wind direction for anemophilous species, should be addressed for established orchards (Sanchez-Estrada and Cuevas, 2020). Limited ability to experiment with new pollinizers.	Established orchards have the chance to enhance pollination through the use of newly developed, compatible pollinizers that may enhance fruit quality and size (for example, through the effects of xenia and metaxenia) (FAO, 2018). For newly established orchards, reducing or eliminating pollinizer, which frequently produces fruit with a lower commercial value than the primary cultivar and may pose a concern if it is more susceptible to pest or disease attack.

It is vital to investigate alternate strategies to improve pollination efficacy in light of these difficulties. Some suggested strategies to mitigate the risks associated with reliance on managed honey bees consist of alternative pollinators (Rader *et al.*, 2020), breeding crops with parthenocarpic traits or greater self-compatibility and self-fertility, thereby reducing dependence on biotic pollination (Knapp *et al.*, 2017). Additionally, manual pollination is a common technique for some high-value species, despite being expensive and labor-intensive. For example, the lack of specialist pollinators outside of Central and South America means that manual pollination is required in the case of vanilla (Hu *et al.*, 2019). On the other hand, mechanical pollination (MP) is a more economically viable method. Compared to natural pollen transfer systems, MP may have the following benefits: consistent pollination rates, less vulnerability to unfavorable environmental factors and climatic fluctuations, and less dependence on currently available insect pollination services (Figure 6.1).

Mechanical pollination (MP) or robotic pollination refers to any type of pollination in which pre-collected pollen is applied to target crops using a mechanical device. On the other hand, for the sake of clarity, non-mechanical pollination systems, also referred to as assisted, controlled, or supplementary pollination are classified as artificial pollination. An extensive overview of artificial pollination in tree crop production was given by Pinillos and Cuevas in 2008(Pinillos and Cuevas, 2008). As a result, the review concentrates on gains made after 2008, paying particular emphasis to MP systems and, when necessary, incorporating knowledge from advances in artificial pollination research. Furthermore, over the past fifteen years, our understanding of pollination biology has significantly increased, especially for newly emerging tree crops like avocado and macadamia (da Silva-Santos *et al.*, 2020), as well as even for well-established tree crops like olives (Montemurro *et al.*, 2019).

Although pollinators play an important role, the focus of research is still on MP systems and their possible uses in commercially important fruit and nut crops. It is suggested that a detailed understanding of the biology of pollination in tree crops might be used to evaluate whether or not MP systems are feasible to adopt. This chapter addresses associated hazards and provides a critical analysis of important aspects of robotic or mechanical pollination systems, including pollen collection, handling, storage, and delivery. The potential for creating MP systems for economically significant fruit and nut crops in the future has also been discussed.

6.2 Advantages of Robotic Pollination Systems

A recent list of researches that have looked into robotic pollination or MP systems since 2008 can be seen in Table 6.2. All eight tree crops (except pistachio) showed a significant increase in fruit and nut set when MP was applied (Table 6.2). Since most of these research used handheld sprayers to apply pollen, it appears that MP systems are still in their early stages of development. Nonetheless, the "grey" sources (Whiting, 2019)make it evident that substantial technological and engineering advancements have been made in the use of MP systems, and that MP is a commercial reality for certain crops, such as date (Mu *et al.*, 2018), sweet cherry (Whiting, 2019), and kiwifruit.For instance, specialized MP equipment for date palm is sold commercially (e.g., AgroPalm Machinery) in New Zealand, several businesses sell machinery to process, extract, dry, store, and distribute kiwifruit pollen (e.g., Fraser Gear) or supply harvested pollen for the kiwifruit industry (e.g., Kiwifruit Vine Health) (Eyles *et al.*, 2022).

Table 6.2: Illustrations of robotic pollination in fruit trees

Crop	Type of application	Pollen information	Pollen carrier	Impact	Reference
Date palm (*Phoenix dactylifera*)	Handheld sprayer	Air dried. Concentrations (0.5, 1.0, 1.5, 2 g L^{-1})	Liquid carrier	7% and 18% more fruit set (measured 7-8 weeks after pollination)	Awad, 2010
Kiwifruit (*Actinidia deliciosa*)	Handheld blowers, tractor-mounted sprayer and hand pollination	Concentration 600 to 1200 g ha^{-1}	Dry (*Lycopo-dium*) and liquid (Pollen Aid) carriers	Pollen application timing is more significant than pollen quantity. Up to 12% more yield was produced.	Asteggiano *et al.*, 2011

Japanese pear (*Pyrus pyrifolia*)	Applying directly by hand and using an electromotive sprayer	Stored at -30°C after being refined with acetone. 35–45% germination	Liquid carrier	Despite inconsistent results, the first fruit set (60–86%) using a liquid carrier was either equal to or less than that of hand pollination (74–94%)	Sakamoto *et al.*, 2009
Olive (*Olea europaea*)	Two times with a mechanical blower	Kept for a few days at 4°C. 35–68% germination	Carrier was not used	The MP fruit set (15%) was more than the control's (3.7%)	Fuxi *et al.*, 2019
Sweet cherry (*Prunus avium*)	Airblast and electrostatic sprayers. Two uses	Concentrations 36, 72, 144 g ha^{-1}	Liquid carrier	Ten studies yielded extremely varied outcomes, ranging from 0-20% increase in fruit set, depending on the cultivar and year	Whiting, 2019
Hazelnut (*Corylus avellana*)	Mist blower	Stored at −20 ◦C. Concen-trations 150 g ha^{-1}	Liquid carrier	Evaluated stigma receptivity and ovule health instead of fruit set	Guastella *et al.*, 2020
Pistachio (*Pistacia vera*)	Three applications with a Handheld sprayer	Nothing specific about the storage settings	Liquid carrier	Inconsistent reactions, although pollen addition did not always result with a better mature fruit set	Karimi *et al.*, 2017

Novel pollination techniques, including mechanical pollination (MP) systems, have the capacity to enhance existing pollination services when implemented in their most basic form, and further benefits can be obtained from more advanced versions (Figure 6.1). The level of complexity that an MP system must have depends on the particular crop's production system. For example, the implementation of an MP system does not always mean that controlled pollinators must be excluded. Alternatively, hybrid MP systems can be used, in which the traditional insect-mediated pollination services are enhanced by applying MP directly or by having insects re-dispense MP-applied pollen, as is already the case in kiwifruit farming. In the long run, a fully functional MP system would include completely replacing pollinizer (pollen donor) trees in commercial orchards with groves planted specifically for the purpose of producing pollen. In order to maximize flower and pollen productivity, orchards specifically designed for the production of sweet cherry pollen were planted in the USA in 2017–2018. Here, specialized pruning and training techniques are being used (Whiting, 2019).

In addition to reducing reliance on controlled pollinators, MP systems provide strategic decision-makers with freedom when it comes to the genetic characteristics of the pollen they apply. Specifically, it has been reviewed that the pollen may have favorable xenic and meta-xenic effects on fruit and nut characteristics, including size, shape, color, flavor, nutrient composition, and ripening dynamics (Yang *et al.*, 2020). For example, it has been demonstrated that xenia effects in hazelnuts can increase nut weight by as much as 45% for one of the four pollinizers under study (Fattahi*et al.*, 2014). Similar to this, it was discovered that cross-pollination with *Murcotttangor* pollen increased the antioxidant profile of Qicheng52 navel oranges by increasing total soluble solids and phenolic content (Wang *et al.*, 2019).

6.3 Procedure of Robotic Pollination

There are usually four essential processes in robotic or mechanical pollination(Cacioppo *et al.*, 2017):1) Pollen collection, 2) Pollen handling and drying 3) Pollen storage and 4) pollen delivery (Figure 7.2).

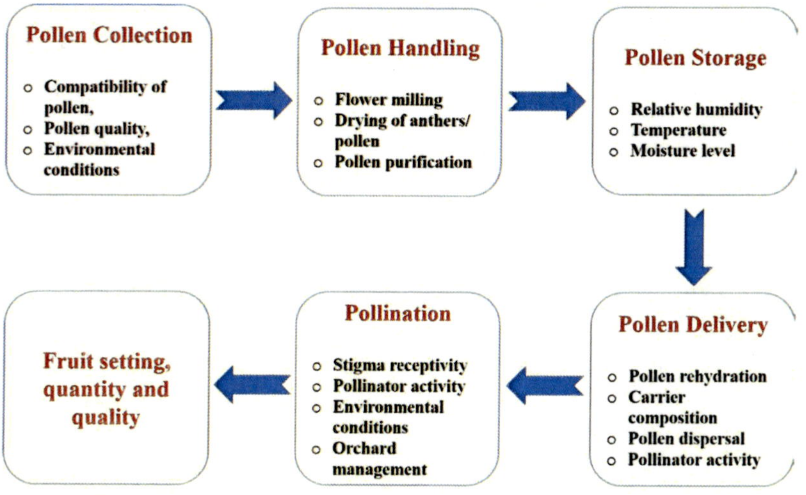

Fig. 6.2: Main components of mechanical pollination system

6.3.1 Pollen Collection

A mechanical pollination (MP) system's ability to function effectively depends on the efficient collecting of high-quality pollen. In order to reduce loss from pollen shedding and avoid contamination, pollen harvesting usually takes place before anthesis (Figure 6.2). Numerous factors, including as genotype (Whiting, 2019), flower maturity (Yi *et al.*, 2006), weather during flowering (Santiago and Sharkey, 2019), and agronomic practices like pesticide administration (Kargar and Imani, 2011), might affect the quality of pollen. For example, in vitro germination rates of nine different pear cultivars (*Pyruscommunis*) cultivated in the same orchard were found to range from 3% to 71% in a research (Cerovic *et al.*, 2020). The next year, these rates increased to 12% to 88%, which was ascribed to warmer temperatures. A minimum of 80% in vitro germination is guaranteed by certain kiwifruit pollen suppliers (Pollen Plus). Although achieving such high germination rates is ideal for maximizing the effectiveness of MP systems, not all tree crops may be able to achieve them or even require them. Indeed, even with pollen exhibiting germination rates of 7.5% and 34.5%, an early hazelnut study showed better fruit set (Ascari *et al.*, 2016).

The accepted theory has continued to hold that bees gather pollen from the bee corbiculae using traps set up at hive entrances. Recent research, however, suggests that pollen gathered by non-corbiculate bees (*Megachilerotundata* and *Halictus* spp.) has more functionality than pollen collected by corbiculate bees (*Bombus impatiens* and *Apismellifera*) (Parker *et al.*, 2015). Anemophilous plants that provide a lot of readily released pollen have effective mechanized harvesting techniques developed for them. For example, it has been claimed that, in optimal shedding conditions, a backpack-mounted pollen vacuum may gather over 500 cc of Douglas fir (*Pseudotsugamenziesii*) pollen every hour (Oster and DeBell, 2016). Similar to this, a modified vacuum approach was tested for gathering significant volumes of pollen from another anemophilous plant, cannabis (*Cannabis sativa*) (Wizenberg *et al.*, 2020).

However, mechanical pollen collection from entomophilous species is still difficult since these species produce less sticky pollen in smaller amounts (Vaknin *et al.*, 1999). As a result, these species' pollen collecting still relies on expensive manual techniques, which may hinder the use of MP systems. However, the existence of multiple commercial pollen suppliers implies that human pollen gathering is still economically feasible, especially in light of MP systems' support for high-value tree commodities.

6.3.2 Pollen Handling

Generally, handling pollen entails physically removing anthers from flowers using methods like milling and/or sieving, then desiccating the anthers and pollen. The last step usually involves gathering cleaned pollen with a suction cyclone device (like Fraser Gear). There are many grades of pollen that can be obtained, from pure to less pure, depending on the requirements for pollen application that must be met in order to obtain the required level of pollination and subsequent fruit/seed set. A range of purity grades are available from some pollen vendors; lower purity grades are usually offered at a discount (e.g., Firman Pollen).Significant progress has been made in pollen handling engineering since 2008. Commercial companies are now selling specialist equipment designed to carry out these tasks for tree crop species that are both entomophilous and anemophilous.

6.3.3 Pollen Storage

Systems for mechanical pollination (MP) rely on using stored pollen. Commercial pollen suppliers claim that, under the right conditions of processing and management, their products can remain viable for two to five years without experiencing noticeable deterioration. There is a good deal of information available regarding pollen long-term storage, mainly due to its critical role in breeding programs and genetic conservation efforts. A large number of research works have examined the impact of temperature on pollen viability over both short- and long-term storage times.

In general, flower pollen kept at 4°C for one to three months is viable, though this will vary according on the type of fruit tree. Pollen lifetime must be extended beyond one to three months by storing it at lower temperatures, such as −20°C. The function of drying in maintaining pollen viability over extended storage has received little research attention (Garcia-Cruzatty *et al.*, 2015). Pollen quality can be affected by delays in the drying process, which can ultimately result in viability loss. Before dispersing, pollen in its natural environment goes through regulated dehydration in the anther. Pollen that has been partially dehydrated has less metabolic activity, which increases its lifetime and resistance to additional desiccation during dispersal (Williams and Brown, 2018).It has been noted that different species have different ideal drying parameters (humidity, temperature, air flow rate, and time) for lowering pollen moisture content. According to some research, 15% is the ideal moisture content; larger amounts (over 30%) are linked to faster degradation (Buitink *et al.*, 1998). High moisture content pollen cells can't survive freezing storage, probably because intracellular ice forms and then membranes rupture (Shivanna and Johri, 1985).However, there seems to be a moisture content threshold below which longevity is not improved and may even jeopardize pollen viability (Buitink *et al.*, 1998).

The number of cells in the grain during maturity, or anthesis, can be used to categorize pollen (Pinillos and Cuevas, 2008). Thirty percent of blooming species produce tricellular pollen, while over seventy percent of them show bicellular pollen (Williams *et al.*, 2014). Fruit tree crops and nut tree crops generally produce bicellular pollen, which is more resistant to drying than tricellular pollen (Brewbaker, 1967). As such, this kind of pollen is easier to store than trinucleate pollen, which is more metabolically active, more differentiated, and has a shorter lifespan than binucleate pollen (Williams and Brown, 2018). As illustrated by the case study of *Nothofagusalpina* (Garcia-Cruzatty *et al.*, 2015), there is, however, a lack of specific knowledge regarding the necessary moisture level to preserve pollen lifetime during storage. This may be because it is difficult to measure moisture content precisely in such minute quantities. Rapid techniques using near-infrared (NIR) technology have been used to measure the amount of moisture in maize and rice seeds (Zhang and Guo, 2020). We argue that precise measurement of the moisture content of pollen is essential to provide a quick indicator of possible pollen quality, especially for large quantities, before long-term preservation.

6.3.4 Pollen Delivery

Many systems for pollen delivery are already available or in development (Table 7.3). The most basic method of achieving mechanical pollination (MP) is to use pollen dispensers, which are placed at the entrance of honey bee or bumble bee colonies and carry pre-collected suitable pollen. When bees leave the hive, they pass through the pollen dispensers and become covered in pollen, which they then scatter on flowers they choose to forage for. Improved dispenser designs are intended to reduce the amount of pollen wasted and make it easier for biological agents to be distributed in order to prevent fungal diseases like *Sclerotinia* and *Botrytis*(Lahdenpera, 2020). Pollen dispensers are useful and cost-effective, but they don't reduce the risk that comes with reliance.

Table 6.3: An overview of pollen delivery methods

Pollen delivery method	Carrier	Remarks	References
Manual (with brush, glass rod, feather etc.)	No	Time consuming and laborious	El Kadriand Ben Mimoun, 2020
Pollen-containing dispenser (found at the beehive's entrance)	Optional (dry)	Reliance on bee activity; low pollinator activity results from bad weather. Enables delivery that is targeted.	Traynor *et al.*, 2020
Handheld sprayer	Optional (dry and wet)	Pollen can be redistributed by bees if a dry carrier is used.	Awad, 2010
Mobile sprayers	Optional (dry and wet)	Pollen can be redistributed by bees if a dry carrier is used.	Ellena *et al.*, 2014
Pollen application sprayers that have booms and a pressure vessel	Optional (dry and wet)	Does not function when it's raining.	Whiting, 2019
Unmanned aerial vehicles - Drones	Optional (dry)	The technology is new. Its pollen-transporting capability is limited (e.g., 5 kg). Does not function when it's raining.	Zhang *et al.*, 2019
Robotics (platform mounted manifold spray nozzle; robotic bees)	Optional (dry and wet)	Precise, independent operation. Technology is a recent development.	Chechetka *et al.*, 2017

Due to the small scale of research projects, most peer-reviewed studies have mostly concentrated on analyzing handheld sprayers/blowers to determine the feasibility of mechanical pollination (MP) systems. But MP systems also need to be assessed at the orchard level using truck-mounted boom sprayers, as the grey literature for almonds, kiwis, and sweet cherries documents (Eyles *et al.*, 2022). A novel mechatronic prototype was tested on kiwifruit with encouraging results. The robotic system showed an 80% spraying accuracy on flowers and was made up of computer-controlled electromechanical components. It was also able to process sensory data and make decisions in real-time, such as flower detection and autonomous operation (Duke *et al.*, 2017).

A common issue with a lot of pollen distribution techniques is that they are not very accurate or efficient. Goodwin and McBrydie report that when pollen is dispersed by traditional broadcast methods, 99% or more of the pollen does not make it to the tiny stigmatic surface (Goodwin and McBrydie, 2013).

Techniques to increase pollen consumption, reduce pollen dosage, and/or improve delivery accuracy have been developed. For example, the use of honey bees to help disperse deposited pollen onto pistillate flowers improved the effectiveness of air delivery techniques used in kiwifruit farming. On the other hand, the use of electrostatic forces has increased the efficacy of wet spray pollination techniques, as charged pollen grains settle on stigmas more readily than uncharged ones (Badger *et al.*, 2015).

When blooms of a trees are only partially responsive when applied, it has been shown that applying more than once is required to improve fruit set. In these cases, the benefits of MP can be offset by the comparatively high price of pollen. According to a recent comparative study comparing the pollination requirements of two kiwifruit cultivars, Zesy002 required almost 66% less pollen grains than Hayward in order to accomplish fruit set (Broussard *et al.*, 2021). These results point to the possibility of customizing MP systems to maximize cost reductions and improve pollen application effectiveness. In addition, field tests using a boom pollen delivery technique showed that a single spray could improve sweet cherry fruit set. However, the timing of application proved crucial, requiring synchronization with a sizable population of newly opened flowers in order to achieve yields that could be sold commercially. Notably, no benefits in fruit set were seen when pollen treatment coincided with rainfall, which highlights the ongoing influence of environmental circumstances on MP systems (Whiting, 2019).

6.4 Possible Hazards Associated with Mechanical Pollination

6.4.1 Pollen Risks

The primary feature of mechanical pollination (MP), which is the collection and transportation of pollen, requires a careful investigation of pollen's possible role as a disease vector. Among the pathogens that are carried by pollen, viruses and viroids are most dangerous. Though there are about 100 species of bacteria known to cause disease to plants, only three species viz., *Xanthomonas arboricola* pv. *juglandis*, *Erwinia amylovora*, and *Pseudomonas syringae* have been linked to pollen-mediated transmission in perennial crops (Donati *et al.*, 2018). However, until recently, direct pathogen detection in field trials was less common than the experimental settings using artificial inoculations that were used to support the role of pollen and pollination in the spread of these infections (Tontou *et al.*, 2014). Donati *et al.* (2018) used a green fluorescent protein mutant to show that *Pseudomonas syringae* pv. *actinidiae* could colonize kiwifruit anthers both endophytically and epiphytically, producing contaminated pollen that could spread the pathogen to healthy plants. Though conjectural, it is still conceivable that *Pseudomonas syringe* pv. *actinidiae* may have entered New Zealand via pollen, causing significant harm to the kiwifruit sector (Vanneste, 2017).

At least 46 of the more than 1000 plant viruses that the International Committee on Taxonomy of Viruses has cataloged have been shown to be spread via pollen (Isogai *et al.*, 2020). A number of these viruses have been identified in hosts that are perennial fruit crops, such as plants belonging to the Vaccinium, Prunus, and Rubus genera (Card *et al.*, 2007). Viral transmission can happen through the produced seed (vertical transmission), as in the case of Prune dwarf virus, or through the pollinated flower (horizontal transmission), as demonstrated by Raspberry bushy dwarf virus (Beaver-Kanuya and Harper, 2019). The commercial movement of pollen across orchards, zones, and countries has significant biosecurity hazards. Therefore, it is critical to create quality assurance processes to guarantee traceability, purity, and absence of pathogenic contamination. In order to detect and quantify viruses and enable the adoption of biosecurity measures, reliable tests with high sensitivity and species specificity like the RT-qPCR assay must be deployed (Beaver-Kanuya and Harper, 2019). Orchardists may further reduce the chance of disease spread by collecting their own pollen, a service provided by a few pollen vendors.

6.4.2 Decreasing Quality and /or Yield

From the limited body of research on mechanical pollination (MP) systems, a pattern of increased crop output is found in most of the investigations. But for some trees crops, more fruit or nut set might not always be a good thing because there could be a negative effect on the quantity and quality of the crop due to resource limitations (Klein *et al.*, 2015). For example, as per research studies increased pollination success could reduce macadamia nut size but not quality (Herbert *et al.*, 2019).

It might be suggested that high-value arboreal crops may have less resource availability barriers due to their increased access to fertilizer inputs. However, an overly abundant fruit set may still have negative consequences for some tree crops. Specifically, when it comes to tree fruit crops where fruit size is important, the use of a mechanical pollination (MP) system requires caution to prevent unintentional increases in thinning costs, especially when it comes to cultivars with high crop loads (Webber *et al.*, 2020). In order to control crop load, maintain marketable fruit sizes and quality, reduce the risk of branch breaking from heavy fruit loads, and avoid biennial bearing events, it is normal to use mechanical and chemical thinning techniques (Webber *et al.*, 2020). To address the latter issue, trees that display tendencies to bear fruit every two years may benefit from the management of those trees by reducing any pollination deficit during off-years, which could potentially disrupt the biennial bearing cycle. This can be achieved by applying mechanical pollination (MP) sparingly during the non-bearing year.

6.5 Elements to Take into Account When Estimating Success of MP for Fruit and Nut Crops

The development of an effective mechanical pollination (MP) system depends on a thorough understanding of the biology of pollination that is unique to the target species. This requires a sophisticated knowledge of the phenology of male and female blooming in various cultivars, which is relevant to the gathering, processing, and storage of pollen as well as efficient methods for delivering pollen that take into account the frequency and timing of pollination episodes. Even cultivars or kinds designated as self-fertile do not rule out the potential use of MP, despite the fact that most tree crops experience cross-pollination. Research, for example, has shown that the fruit set of self-fertile almond cultivars increased by 60% in bee-pollinated trees (20%) as compared to specimens that were bee-isolated (31%), according to studies (Saez *et al.*, 2020). In the same way, *Coffee arabica*, which is self-pollinating, shows a 35% increase in output (by weight) when bees visit the blooms (Roubik, 2002), highlighting the practicality of MP systems in such crops. Moreover, due to global demand or comparative advantage, the majority of tree crops are farmed outside of their original geographic limits. As an instance, Brazil is currently the eighth-largest producer of macadamias in the world. However, research on the biology of pollination of macadamia crops in Brazil is said to be limited and in its early stages. It was found that butterflies are the main floral visitors (they account for half of all floral visits) (da Silva-Santos *et al.*, 2020), which is in contrast to observations of native stingless bees and exotic honey bees that are common in the native Australian habitat (Howlett *et al.*, 2015). Therefore, a thorough grasp of pollination biology becomes crucial for determining if MP is feasible for a certain tree crop. Important factors include traits of the pollen, flower form, flowering times, and stigma receptivity, which are explained below.

6.5.1 Type of Pollen

Binucleate pollen is present in the fruit and nut crops, which indicates that the crops are viable after drying and can be stored for an extended period of time. However, pollen can be divided into two categories according to its water content at dispersal: orthodox and recalcitrant forms, with the former being somewhat dried (20%) and the latter being totally hydrated, as outlined by Pacini and Dolferus (Pacini and Dolferus, 2019). In practice, this means that orthodox pollen—which is more resistant to dehydration might be easier to handle than recalcitrant pollen, which is more sensitive to dehydration. When it comes to tree crops with recalcitrant pollen, like avocado, hazelnut, mango, pecan, pistachio, and walnut, developing procedures to maintain pollen viability could be more difficult. In fact, post-storage germination rates for mango, pistachio, and pecan have significantly decreased, according to pollen storage trials; for mango, pollen viability is maintained only at -196°C.

6.5.2 Morphology of Flower

The correct placement of pollen on the stigma is essential to the effectiveness of mechanical pollination (MP), yet current delivery techniques frequently fail to achieve this. To maximize the likelihood of pollen adherence, the stigma should ideally be completely exposed and free of any physical obstructions. With a few notable exceptions, a careful study of floral morphology indicates that most tree crops meet this requirement. The Brazil nut flower is especially notable because it has an androecial hood that is usually unfurled by the pollinator to gain access to the stigma (Cavalcante *et al.*, 2012). Furthermore, the style of the blueberry flower is sometimes much shorter than the urn-shaped corolla, which makes it difficult for pollen carried by the wind to efficiently reach the stigma.

6.5.3 Pattern of Flowering and Receptivity of Stigma

Important information required to establish the best time and frequency of mechanical pollination (MP) treatments includes temporal flowering patterns at the tree level and stigma receptivity at the flower level. Furthermore, additional variations in these characteristics might occur within a specific species, cultivar, or site, as well as in other environmental circumstances (von Bennewitz *et al.*, 2019). Considering this variability, it is wiser to link stigma receptivity mainly to the developmental stage instead of following a strict chronological sequence (Zhang *et al.*, 2015). The technical difficulties in applying MP commercially become evident for tree crops with extremely short periods of stigma receptivity, as some avocado cultivars that flower female in the afternoon and male in the morning. Kiwifruit, on the other hand, has the same ability to produce fruit at any time of day, providing flexibility to use MP outside of the typical hours of pollinator activity (Broussard *et al.*, 2021).

When it comes to dichogamous tree crops that is, protandrous cultivars whose pollen sheds before stigma receptivity, there is a chance to collect pollen within the same growing season and use fresh pollen instead of storing it. For example, early in the season, up to two weeks before stigma receptivity, some male cultivars of pecans discharge pollen (Zhang *et al.*, 2015). Significantly, hazelnut stigmas in their natural environment are receptive for several months;

on the other hand, because of warmer and drier winter circumstances, hazelnuts grown outside of their customary locations have far less receptivity, lasting only one or two weeks (Guastella *et al.*, 2020). It would seem that the creation of MP systems requires customized strategies that take into account regional agriculture systems and current environmental circumstances.

6.5.4 Training System

Fruit orchard management techniques have changed significantly over the last 50 years, moving from conventional vase training systems to greater density trellised two-dimensional (2-D) training systems. It has been demonstrated that integrating these systems with dwarfing rootstocks improves early light interception and makes it easier for light to enter the canopy (Dallabetta *et al.*, 2021), opening the way for later automation improvements that are expected, especially in mechanical pruning and harvesting processes. Additionally, recent studies have looked into whether using robotic pollination techniques is feasible. Adopting contemporary orchard methods has benefits for canopy management in addition to improving the effectiveness of spray applications, such as mechanical pollination (MP) systems.

Since 2-D systems have a compact and thin canopy, the floral canopy is more uniform and easily accessible, which helps pollen disperse evenly and lowers the amount of pollen needed. Findings from functional structural plant modeling (FSPM) may be useful in the creation and maintenance of canopy structures optimized for pollen transport. The development of thorough crop models that take into consideration complex elements including phenological events, architectural dynamics, and physiological processes at the tree level has been made easier by advancements in FSPM (Louarn and Song, 2020). Mango and macadamia are two examples of tree crops whose growth and fruit production dynamics have been successfully studied by FSPM (Auzmendi and Hanan, 2020). For example, the macadamia model concentrated on replicating patterns of carbon allocation, whereas the mango model simulated the formation of growth units and inflorescences.

6.5.5 Tree Height

Certain fruit and nut trees, including macadamia and Brazil nuts, can grow to heights of up to 30 to 50 meters, especially in situations when dwarfing rootstocks are unavailable (Sales *et al.*, 2021). It's obvious that these crops have difficulties receiving pollen from the ground. A US-based company has realized the commercial use of drones for aerial pollen delivery for crops like apple, almond, sweet cherry, and pear; a system for dates (23 m tall) is presently under development (Dropcopter). Due to limited carrying capacity

(5 kg) in drone pollination, most pollen is transported without a carrier. It is uncertain if drone pollination would be a practical solution for these tall tree crops, but it might only be possible on windless days, which could have an effect on the MP systems' capacity to make a profit.

6.6 Conclusion

Pinillos and Cuevas (2008) raised the query if robotic or mechanical pollination (MP) could replace natural pollination as this would allow for the application and selection of the best pollen at the right time to guarantee adequate yields. Since then, MP systems have mostly served as supplements to natural pollination processes rather than as a substitute, even if they are now commercially available for some fruit and nut crops. Nevertheless, a number of obstacles stand in the way of MP systems' broad adoption and development. In the first place, there are serious gaps in our knowledge of the biology of fruit and nut pollination that affect economies. The creation of MP systems is hampered by these knowledge gaps, especially for newly developing crop types that are farmed outside of their natural areas. Furthermore, little is known about the effects of human-caused climate change on plant phenology and flower development, and changes in pollinator distribution patterns.Second, empirical results are primarily at the tree scale in peer-reviewed literature, which reflects the pragmatic limitations of research projects. Scaling up to demonstrate the cost-effectiveness and logistical feasibility in relation to improved pollination results across spatial and temporal dimensions is necessary to persuade the horticultural industry of MP systems' efficacy.

Thirdly, consideration must be given to the function of wild pollinator species and the increasing use of non-Apis species in almond pollination, including bumblebees and solitaryOsmia bees. There are still questions about how they might affect MP systems' efficacy, as they were primarily developed to supplement European honey bee use.In final, the presence of companies providing premium pollen, specialist MP equipment, and MP services highlights a possible discrepancy between results published in literature and proprietary commercial information originating from internal R&D initiatives. One of the obstacles to MP systems' growth into other agricultural fields is limited access to this exclusive expertise. MP systems are designed to compensate for pollination deficits in highly intensive, risk-averse sectors. Following stakeholder assurances about the reliable operation and financial sustainability of these technologies and systems, quick investment and uptake of MP are expected to reduce the risks related to below-average agricultural yields and quality resulting from insufficient pollination.

References

Ascari, L., Cristofori, V., Macrì, F., Botta, R., Silvestri, C., De Gregorio, T., Huerta, E. S., Di Berardino, M., Kaufmann, S., &Siniscalco, C. (2020). Hazelnut pollen phenotyping using label-free impedance flow cytometry. Frontiers in Plant Science, 11, 615922.

Ascari, L., Guastella, D., Sigwebela, M., Engelbrecht, G., Stubbs, O., Hills, D., De Gregorio, T., &Siniscalco, C. (2017). Artificial pollination on hazelnut in South Africa: preliminary data and perspectives. Acta Horticulturae,1226, 141-148.

Asteggiano, L., Giordani, L., Bevilacqua, A., Vittone, G., & Costa, G. (2011). Ten years of research on complementary pollination of kiwifruit. Acta Horticulturae, 913, 615-620.

Auzmendi, I., &Hanan, J. S. (2020). Investigating tree and fruit growth through functional–structural modelling: implications of carbon autonomy at different scales. Annals of Botany, 126(4), 775-788.

Awad, M. A. (2010). Pollination of date palm (Phoenix dactylifera L.) cv. Khenazy by pollen grain-water suspension spray. Journal of Food, Agriculture & Environment, 8, 313-317.

Badger, M., Ortega-Jimenez, V. M., von Rabenau, L., Smiley, A., & Dudley, R. (2015). Electrostatic charge on flying hummingbirds and its potential role in pollination. PLoS One, 10(9), e0138003.

Beaver-Kanuya, E., & Harper, S. J. (2019). Detection and quantification of four viruses in Prunus pollen: Implications for biosecurity. Journal of Virological Methods, 271, 113673.

Bosch, J., Osorio-Canadas, S., Sgolastra, F., &Vicens, N. (2021). Use of a managed solitary bee to pollinate almonds: Population sustainability and increased fruit set. Insects, 12(1), 56.

Brewbaker, J. L. (1967). The distribution and phylogenetic significance of binucleate and trinucleate pollen grains in the angiosperms. American Journal of Botany, 54(9), 1069-1083.

Broussard, M. A., Goodwin, M., McBrydie, H. M., Evans, L. J., &Pattemore, D. E. (2021). Pollination requirements of kiwifruit (Actinidiachinensis Planch.) differ between cultivars 'Hayward'and 'Zesy002'. New Zealand Journal of Crop and Horticultural Science, 49(1), 30-40.

Buitink, J., Claessens, M. M., Hemminga, M. A., & Hoekstra, F. A. (1998). Influence of water content and temperature on molecular mobility and intracellular glasses in seeds and pollen. Plant Physiology, 118(2), 531-541.

Cacioppo, O., Michelotti, V., Vittone, G., &Tacconi, G. (2017). Pollination of kiwifruit: 30 years of applied research leads to a model system for studying the interaction between pollination and flowering stage. Acta Horticulturae,1229, 355–363.

Card, S. D., Pearson, M. N., & Clover, G. R. G. (2007). Plant pathogens transmitted by pollen. Australasian Plant Pathology, 36, 455-461.

Cavalcante, M. C., Oliveira, F. F., Maues, M. M., &Freitas, B. M. (2012). Pollination requirements and the foraging behavior of potential pollinators of cultivated Brazil nut (BertholletiaexcelsaBonpl.) trees in central Amazon rainforest. Psyche: a Journal of Entomology, 978019.

Cerovic, R., FotiricAksic, M., &Meland, M. (2020). Success rate of individual pollinizers for the pear cultivars "Ingeborg" and "Celina" in a Nordic climate. Agronomy, 10(7), 970.

Chechetka, S. A., Yu, Y., Tange, M., &Miyako, E. (2017). Materially engineered artificial pollinators. Chem, 2(2), 224-239.

Dallabetta, N., Guerra, A., Pasqualini, J., & Fazio, G. (2021). Performance of semi-dwarf apple rootstocks in two-dimensional training systems. HortScience, 56(2), 234-241.

da Silva-Santos, R., de Oliveira Milfont, M., Silva, M. M., Carneiro, L. T., & Castro, C. C. (2020). Butterflies provide pollination services to macadamia in northeastern Brazil. ScientiaHorticulturae, 259, 108818.

Donati, I., Cellini, A., Buriani, G., Mauri, S., Kay, C., Tacconi, G., &Spinelli, F. (2018). Pathways of flower infection and pollen-mediated dispersion of Pseudomonas syringaepv. actinidiae, the causal agent of kiwifruit bacterial canker. Horticulture Research, 5, 56.

Duke, M., Barnett, J., Bell, J., Jones, M. H., Martinson, P., McDonald, B., & Lim, J. (2017). Automated pollination of kiwifruit flowers. In 7th Asian-Australasian Conference on Precision Agriculture (pp. 1-5).

El Kadri, N., & Ben Mimoun, M. (2020). In vitro germination of different date palm (Phoenix dactylifera L.) pollen sources from Southern Tunisia under the effect of three storage temperatures. International journal of fruit science, 20(sup3), S1519-S1529.

Ellena, M., Sandoval, P., Gonzalez, A., Galdames, R., Jequier, J., Contreras, M., &Azocar, G. (2014). Preliminary results of supplementary pollination on hazelnut in south Chile. Acta Horticulturae, 1052, 121–128.

Eyles, A., Close, D. C., Quarrell, S. R., Allen, G. R., Spurr, C. J., Barry, K. M., Whiting, M. D., & Gracie, A. J. (2022). Feasibility of mechanical pollination in tree fruit and nut crops: a review. Agronomy, 12(5), 1113.

FAO. (2018). The Pollination of Cultivated Plants: A Compendium for Practitioners. Roubik, D. W., Ed. FAO: Rome, Italy, Volume 2.

Fattahi, R., Mohammadzedeh, M., &Khadivi-Khub, A. (2014). Influence of different pollen sources on nut and kernel characteristics of hazelnut. ScientiaHorticulturae, 173, 15-19.

Forbes, S. J., Mustiga, G., Romero, A., Northfield, T. D., Lambert, S., &Motamayor, J. C. (2019). Supplemental and synchronized pollination may increase yield in cacao. HortScience, 54(10), 1718-1727.

Fuxi, S. H. I., Zongjin, J. I. A. N. G., Chi, M. A., &Yue, Z. H. U. (2019). Controlled Parameters of Targeted Pollination Depositionby Air-Liquid Nozzle. NongyeJixieXuebao/ Transactions of the Chinese Society of Agricultural Machinery, 50(12).

Garcia-Cruzatty, L., Rivero, M., &Droppelmann, F. (2015). Effect of temperature and drying on the longevity of stored Nothofagusalpina pollen. New Zealand Journal of Botany, 53(3), 155-164.

Goodrich, B. K., Williams, J. C., & Goodhue, R. E. (2019). The great bee migration: Supply analysis of honey bee colony shipments into California for almond pollination services. American Journal of Agricultural Economics, 101, 1353–1372.

Goodwin, R. M., &McBrydie, H. M. (2013). Use of pollen blowers and pollen dispensers to pollinate kiwifruit artificially. A Report Prepared for. Zespri Group Ltd.

Guastella, D., Sigwebela, M., Suarez, E., Stubbs, O., Acevedo, J., &Engelbrecht, G. (2020). Effect of photo-selective shade nets on pollination process and nut development of Corylusavellana L. Frontiers in Plant Science, 11, 602766.

Guichard, M., Dietemann, V., Neuditschko, M., &Dainat, B. (2020). Advances and perspectives in selecting resistance traits against the parasitic mite Varroa destructor in honey bees. Genetics Selection Evolution, 52, 1-22.

Hatfield, J. L., &Prueger, J. H. (2015). Temperature extremes: Effect on plant growth and development. Weather and Climate Extremes, 10, 4-10.

Herbert, S. W., Walton, D. A., & Wallace, H. M. (2019). The influence of pollen-parent and carbohydrate availability on macadamia yield and nut size. ScientiaHorticulturae, 251, 241-246.

Howlett, B. G., Nelson, W. R., Pattemore, D. E., & Gee, M. (2015). Pollination of macadamia: Review and opportunities for improving yields. ScientiaHorticulturae, 197, 411-419.

Hu, Y., ResendeJr, M. F., Bombarely, A., Brym, M., Bassil, E., & Chambers, A. H. (2019). Genomics-based diversity analysis of Vanilla species using a Vanilla planifolia draft genome and Genotyping-By-Sequencing. Scientific Reports, 9(1), 3416.

Isaacs, R., Williams, N., Ellis, J., Pitts-Singer, T. L., Bommarco, R., & Vaughan, M. (2017). Integrated crop pollination: combining strategies to ensure stable and sustainable yields of pollination-dependent crops. Basic and Applied Ecology, 22, 44-60.

Isogai, M., Miyoshi, K., Watanabe, M., & Yoshikawa, N. (2020). Characterization of horizontal transmission of blueberry latent spherical virus by pollen. Archives of Virology, 165, 2807-2815.

Kargar, M. H., & Imani, A. (2011). Effects of fungicides on pollen germination peach and nectarine in vitro. African Journal of Plant Science, 5(11), 643-647.

Karimi, H. R., Mohammadi, N., Estaji, A., &Esmaeilizadeh, M. (2017). Effect of supplementary pollination using enriched pollen suspension with Zn on fruit set and fruit quality of pistachio. ScientiaHorticulturae, 216, 272-277.

Klein, A. M., Hendrix, S. D., Clough, Y., Scofield, A., &Kremen, C. (2015). Interacting effects of pollination, water and nutrients on fruit tree performance. Plant Biology, 17(1), 201-208.

Knapp, J. L., Bartlett, L. J., & Osborne, J. L. (2017). Re-evaluating strategies for pollinator-dependent crops: How useful is parthenocarpy?. Journal of Applied Ecology, 54(4), 1171-1179.

Kreitzman, M., Toensmeier, E., Chan, K., Smukler, S., &Ramankutty, N. (2020). Perennial staple crops: yields, distribution, and nutrition in the global food system. Frontiers in Sustainable Food Systems, 4, 216.

Lahdenpera, M. L. (2020). A case study: Use of prestop® mix biofungicide in entomovectoring on apple against storage rot diseases. Entomovectoring for Precision Biocontrol and Enhanced Pollination of Crops, 137-145.

Louarn, G., & Song, Y. (2020). Two decades of functional–structural plant modelling: now addressing fundamental questions in systems biology and predictive ecology. Annals of Botany, 126(4), 501-509.

Montemurro, C., Dambruoso, G., Bottalico, G., &Sabetta, W. (2019). Self-incompatibility assessment of some Italian olive genotypes (Oleaeuropaea L.) and cross-derived seedling selection by SSR markers on seed endosperms. Frontiers in Plant Science, 10, 426068.

Mu, L., Liu, H., Cui, Y., Fu, L., &Gejima, Y. (2018). Mechanized technologies for scaffolding cultivation in the kiwifruit industry: A review. Information Processing in Agriculture, 5(4), 401-410.

Oster, R., &DeBell, J. (2016). Development of a backpack-mounted pollen vacuum. Tree Plant. Notes, 59, 42-47.

Pacini, E., &Dolferus, R. (2019). Pollen developmental arrest: maintaining pollen fertility in a world with a changing climate. Frontiers in Plant Science, 10, 429534.

Padilla, F., Soria, N., Oleas, A., Rueda, D., Manjunatha, B., Kundapur, R. R., Maddela, N. R., &Rajeswari, B. (2017). The effects of pesticides on morphology, viability, and germination of Blackberry (RubusglaucusBenth.) and Tree tomato (Solanumbetaceum Cav.) pollen grains. 3 Biotech, 7, 1-12.

Parker, A. J., Tran, J. L., Ison, J. L., Bai, J. D. K., Weis, A. E., & Thomson, J. D. (2015). Pollen packing affects the function of pollen on corbiculate bees but not non-corbiculate bees. Arthropod-Plant Interactions, 9, 197-203.

Pinillos, V., & Cuevas, J. (2008). Artificial pollination in tree crop production. In: Horticultural Reviews. Janick, J., Ed.; Wiley and Sons, Hoboken, NJ, USA, pp. 239–276.

Quinet, M., Mabeluanga, T., Moquet, L., &Jacquemart, A. L. (2016). Introduction of new tools to improve pollination in European pear orchards. ScientiaHorticulturae, 213, 5-12.

Quinlan, G., Milbrath, M., Otto, C., Smart, A., Iwanowicz, D., Cornman, R. S., & Isaacs, R. (2021). Honey bee foraged pollen reveals temporal changes in pollen protein content and changes in forager choice for abundant versus high protein flowers. Agriculture, Ecosystems & Environment, 322, 107645.

Rader, R., Cunningham, S. A., Howlett, B. G., & Inouye, D. W. (2020). Non-bee insects as visitors and pollinators of crops: Biology, ecology, and management. Annual Review of Entomology, 65, 391-407.

Rezaei, M., Hokmabadi, H., Khadivi, A., Safari-Khuzani, A., &Heidari, P. (2019). The selection of superior pistachio (Pistaciavera L.) genotypes among seedling trees originated from open-pollination. ScientiaHorticulturae, 251, 88-100.

Roubik, D. W. (2002). Feral African bees augment neotropical coffee yield. Pollinating bees: the conservation link between agriculture and nature, 255-266.Kevan, P.G., Impereatriz-Fonseca, V., Eds.; Ministry of the Environment: Brasilia, Brazil, pp. 255–266.

Saez, A., Aizen, M. A., Medici, S., Viel, M., Villalobos, E., &Negri, P. (2020). Bees increase crop yield in an alleged pollinator-independent almond variety. Scientific Reports, 10(1), 3177.

Saez, A., Negri, P., Viel, M., &Aizen, M. A. (2019). Pollination efficiency of artificial and bee pollination practices in kiwifruit. ScientiaHorticulturae, 246, 1017-1021.

Sakamoto, D., Hayama, H., Ito, A., Kashimura, Y., Moriguchi, T., & Nakamura, Y. (2009). Spray pollination as a labor-saving pollination system in Japanese pear (Pyruspyrifolia (Burm. f.) Nakai): development of the suspension medium. ScientiaHorticulturae, 119(3), 280-285.

Sales, L. P., Rodrigues, L., &Masiero, R. (2021). Climate change drives spatial mismatch and threatens the biotic interactions of the Brazil nut. Global Ecology and Biogeography, 30(1), 117-127.

Sanchez-Estrada, A., & Cuevas, J. (2020). Pollination designs in 'Manzanillo' olive orchards. ScientiaHorticulturae, 261, 108918.

Santiago, J. P., & Sharkey, T. D. (2019). Pollen development at high temperature and role of carbon and nitrogen metabolites. Plant, Cell & Environment, 42(10), 2759-2775.

Serna. (2020). Available online: https://www.nutfruit.org/industry/technical-resources?category=statistical-yearbooks (accessed on 11 March 2021).

Shivanna, K. R., &Johri, B. M. (1985). The angiosperm pollen: structure and function. Wiley Eastern: Louisville, KY, USA.

Simon, S., Lesueur-Jannoyer, M., Plenet, D., Lauri, P. E., & Le Bellec, F. (2017). Methodology to design agroecological orchards: Learnings from on-station and on-farm experiences. European Journal of Agronomy, 82, 320-330.

Statista. (2020). Available online: https://www.statista.com/statistics/264001/worldwide-production-of-fruit-by-variety/ (accessed on 12 March 2024).

Toledo-Hernandez, M., Wanger, T. C., &Tscharntke, T. (2017). Neglected pollinators: Can enhanced pollination services improve cocoa yields? A review. Agriculture, Ecosystems & Environment, 247, 137-148.

Tontou, R., Giovanardi, D., & Stefani, E. (2014). Pollen as a possible pathway for the dissemination of Pseudomonas syringaepv. actinidiae and bacterial canker of kiwifruit. PhytopathologiaMediterranea, 333-339.

Traynor, K. S., Mondet, F., de Miranda, J. R., Techer, M., Kowallik, V., Oddie, M. A., Chantawannakul, P., & McAfee, A. (2020). Varroa destructor: A complex parasite, crippling honey bees worldwide. Trends in Parasitology, 36(7), 592-606.

Vaknin, Y., Gan-Mor, S., Bechar, A., Ronen, B., &Eisikowitch, D. (1999). Effects of desiccation and dilution on germinability of almond pollen. The Journal of Horticultural Science and Biotechnology, 74(3), 321-327.

Vanneste, J. L. (2017). The scientific, economic, and social impacts of the New Zealand outbreak of bacterial canker of kiwifruit (Pseudomonas syringaepv. actinidiae). Annual Review of Phytopathology, 55, 377-399.

vonBennewitz, E., Ramirez, C., Munoz, D., Cazanga-Solar, R., Losak, T., Alba-Mejia, J. E., &Maureira-Butler, I. (2019). Phenology, pollen synchronization and fruit characteristics of european hazelnut (Corylusavellana L.) cv. tonda de Giffoni in three sites of central Chile. Revista de la Facultad de CienciasAgrariasUNCuyo, 51(2), 55-67.

Wang, Q., Zheng, Y., Yu, Y., Gao, H., Lai, C., Luo, X., & Huang, X. (2019). Effects of cross-pollination by 'Murcott'tangor on the physicochemical properties, bioactive compounds and antioxidant capacities of 'Qicheng 52'navel orange. Food Chemistry, 270, 476-480.

Webber, S. M., Garratt, M. P., Lukac, M., Bailey, A. P., Huxley, T., & Potts, S. G. (2020). Quantifying crop pollinator-dependence and pollination deficits: The effects of experimental scale on yield and quality assessments. Agriculture, Ecosystems & Environment, 304, 107106.

Whiting, M. (2019). Advancing Precision Pollination Systems to Improve Yield Security. Washington Tree Fruit Research Commission: Prosser, WA, USA, p.11.

Williams, J. H., & Brown, C. D. (2018). Pollen has higher water content when dispersed in a tricellular state than in a bicellular state. ActaBotanicaBrasilica, 32, 454-461.

Williams, J. H., Taylor, M. L., & O'Meara, B. C. (2014). Repeated evolution of tricellular (and bicellular) pollen. American Journal of Botany, 101(4), 559-571.

Wizenberg, S. B., Weis, A. E., & Campbell, L. G. (2020). Comparing methods for controlled capture and quantification of pollen in Cannabis sativa. Applications in Plant Sciences, 8(9), e11389.

Yang, Q., Fu, Y., Liu, Y., Zhang, T., Peng, S., & Deng, J. (2020). Novel classification forms for xenia. HortScience, 55(7), 980-987.

Yi, W., Law, S. E., McCoy, D., &Wetzstein, H. Y. (2006). Stigma development and receptivity in almond (Prunusdulcis). Annals of Botany, 97(1), 57-63.

Zhang, C., Valente, J., Kooistra, L., Guo, L., & Wang, W. (2019). Opportunities of UAVs in orchard management. The International Archives of the Photogrammetry, Remote Sensing and Spatial Information Sciences, 42, 673-680.

Zhang, R., Peng, F., Liang, Y., Hao, M., & Li, Y. (2015). Flowering and pollination characteristics of Chinese-grown pecan (Caryaillinoinensis). ActaHorticulturae, 1070(1070), 43-51.

Zhang, Y., &Guo, W. (2020). Moisture content detection of maize seed based on visible/near-infrared and near-infrared hyperspectral imaging technology. International Journal of Food Science and Technology, 55(2), 631-640.

7

Future of AI in Horticultural Crops

Wani Farhana, Gazala Gulzar and Zahoor Ahmad Bhat

Department of Plant Pathology, Sher-e-Kashmir University of Agricultural Sciences & Technology of Kashmir, Shalimar, Srinagar-190025, J&K, India

Abstract

Artificial Intelligence (AI) represents a transformative paradigm in horticulture, offering innovative solutions to enhance productivity, sustainability, and resilience in crop management. This abstract explores the multifaceted applications and future prospects of AI in horticultural crops, including precision agriculture, crop monitoring, genetic improvement, supply chain optimization, and automated harvesting. Key AI technologies, such as machine learning, computer vision, robotics, and natural language processing, are poised to revolutionize crop management practices by enabling real-time decision-making, optimizing resource utilization, and enhancing market intelligence. Despite challenges such as data acquisition, model interpretability, and access to advanced computing resources, initiatives are underway to overcome these hurdles and foster the wider adoption of AI in horticulture. Promising applications, including yield forecasting, disease diagnosis, and fruit harvesting robots, have demonstrated the potential of AI to address critical issues in horticulture, such as market fluctuations and climate variability. Ethical considerations and equitable access to AI benefits must be prioritized as the field progresses to ensure a sustainable and productive agricultural landscape.

***Keywords**: Artificial Intelligence, Horticulture Crops, Crop Monitoring, Automated Harvesting, Yield Forecasting, Disease Diagnosis, Robotics*

7.1 Introduction

Artificial intelligence (AI) is the imitation of human intellect in machines designed to emulate human activities and cognitive functions. These machines are designed to accomplish tasks that normally require human intellect, such as learning, problem solving, perception, and decision-making (Zhang *et al.*, 2022). Horticultural crops, including fruits, vegetables, nuts, and ornamental plants, are of paramount importance in global agriculture, as they contribute significantly to food security, economic prosperity, and environmental sustainability. Nevertheless, the cultivation of these crops poses numerous challenges for growers, such as fluctuating market demand, insufficient resources, unpredictable climate conditions, and the constant threat of pests and diseases (Nowak & Grant, 2018). In recent years, the agricultural sector has experienced a significant shift owing to advancements in Artificial Intelligence (AI) technologies (Liakos *et al.*, 2018). AI has emerged as a powerful tool for the cultivation of horticultural crops, providing solutions for boosting productivity, sustainability, and efficiency (Shrivastava & Kumar, 2021). By utilizing AI, horticulturists can overcome the traditional challenges associated with crop management, resource allocation, and environmental sustainability, paving the way for a more resilient and productive agricultural sector (Garg *et al.*, 2019). The integration of AI into horticultural crop production encompasses a diverse range of applications, including precision agriculture, crop monitoring and management, predictive analytics, automated harvesting, and genetic improvement (Golhani *et al.*, 2018). These applications leverage machine learning algorithms, remote sensing technologies, and data analytics to analyze vast amounts of agricultural data ranging from soil composition and weather patterns to crop health and market trends (Araus& Cairns, 2014). Through intelligent data-driven insights, AI enables farmers to make informed decisions in real-time, optimize resource utilization, minimize waste, and maximize yields (Kamilaris *et al.*, 2017). This technological convergence offers promising prospects for food production, enabling stakeholders to explore new avenues for growth, resilience, and sustainability in horticultural crop cultivation (Mohanty *et al.*, 2016).

7.2 Foundations of Ai Technologies

In the future of horticulture crops, AI technologies will serve as foundational pillars, reshaping how we cultivate, monitor, and optimize crop production Fig.10.1. Here's how these foundational AI technologies will revolutionize horticulture:

1. Machine learning, and deep learning will greatly enhance decision-making in horticulture, as algorithms utilizing vast datasets from sensors, drones, and satellites will be implemented (Smith *et al.*, 2023). Deep learning techniques such as convolutional neural networks (CNNs) will enable the analysis of plant health indicators from images and sensor data, facilitating early disease detection, nutrient deficiency identification, and pest infestation recognition (Jones & Wang, 2022).
2. Natural language processing (NLP) will enable seamless communication between farmers and AI-powered systems, allowing farmers to interact

with smart farming assistants through voice commands or text inputs to receive real-time, tailored insights, recommendations, and alerts (Brown & Garcia, 2024).

3. Computer vision systems will play a crucial role in monitoring and managing horticultural crops, as drones equipped with high-resolution cameras and AI algorithms will patrol fields to assess crop health, detect weeds, and estimate yields, while ground-based cameras will monitor greenhouse environments continuously to ensure optimal growth conditions (Chen *et al.*, 2023).
4. Robotics will revolutionize labour-intensive tasks in horticulture by enhancing efficiency and reducing operational costs, as AI-powered robots equipped with computer vision systems and robotic arms will autonomously perform activities such as planting, pruning, and harvesting with precision and accuracy These robotic systems will operate around the clock, maximizing productivity while minimizing the need for human intervention (Gupta & Lee, 2023).

7.3 Advantages of Ai in Horticulture Crops

In recent years, the integration of artificial intelligence (AI) in agriculture has revolutionized horticultural practices, offering a myriad of benefits to farmers worldwide. From enhancing efficiency in data processing to enabling precise resource management and bolstering resilience against climate change, AI-powered solutions are driving significant advancements in crop cultivation. Fig.10.1. Various ways AI is transforming horticulture, optimizing yields, reducing costs, and fostering sustainability are:

1. **Increased Efficiency**: The AI-powered systems allows farmers to swiftly process copious amounts of data, enabling them to make informed decisions about crop management, including resource allocation, planting schedules, and pest control (Smith *et al.*, 2023). As a result, the utilization of water, fertilizers, and pesticides is optimized, leading to cost reductions and improved crop yields.
2. **Enhanced Crop Health Monitoring**: By employing AI algorithms to analyze images captured by drones or ground-based cameras, farmers can detect early signs of diseases, nutrient deficiencies, or pest infestations (Jones & Wang, 2022). Continuous crop health monitoring enables farmers to intervene promptly to prevent disease spread and minimize yield losses.
3. **Optimized Resource Management**: AI technologies facilitate precision agriculture practices, empowering farmers to tailor irrigation,

fertilization, and pesticide application according to each crop's specific needs and field conditions (Brown & Garcia, 2024). This optimization of resource management conserves water, reduces chemical inputs, and minimizes environmental impact, promoting sustainable agricultural practices.

4. **Improved Crop Breeding and Genetics**: AI-driven genomics and breeding methods have accelerated the development of new crop varieties with desirable features, such as disease resistance, drought tolerance, and enhanced nutritional content (Chen *et al.*, 2023). AI algorithms can analyze genomic data and identify genes associated with specific traits, allowing breeders to make targeted crosses and selections, thus speeding up the breeding process.
5. **Labour Savings and Automation**: Robotics and automation technologies fuelled by AI reduce the need for manual labour in horticulture operations, including planting, pruning, and harvesting (Gupta & Lee, 2023). Equipped with computer vision, autonomous robotic systems can perform tasks with precision and efficiency, resulting in labour savings, increased productivity, and diminished dependence on seasonal labour.
6. **Enhanced Market Intelligence**: AI-powered analytics tools analyze market trends, consumer preferences, and supply chain dynamics to provide farmers with valuable insights into pricing, demand forecasting, and market opportunities (Javaid*et al.*, 2023). By leveraging this information, farmers can make data-driven decisions about which crops to grow, when to harvest, and where to sell their produce, thus maximizing profitability.
7. **Resilience to Climate Change**: AI technologies help farmers adapt to changing environmental conditions and mitigate the impacts of climate change on horticultural crops (Taylor & Smith, 2023). By providing real-time weather forecasts, predictive analytics, and risk assessments, AI enables farmers to anticipate and respond to extreme weather events, water shortages, and other climate-related challenges, thereby ensuring the resilience of agricultural systems.

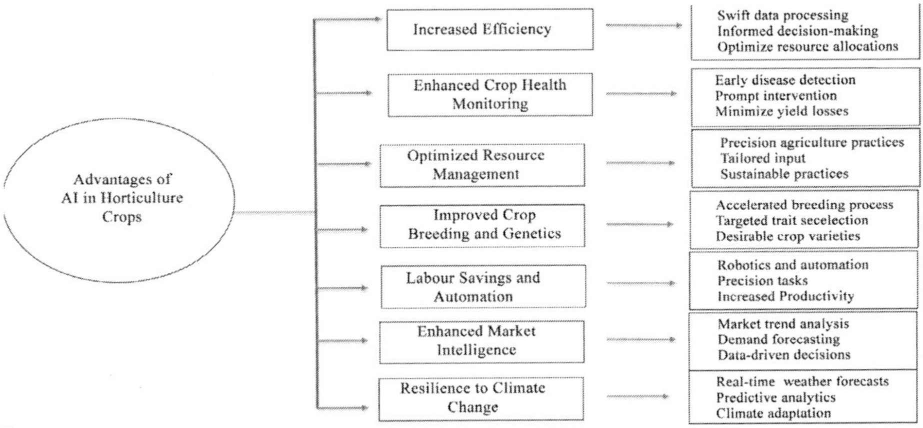

Fig. 7.1: Advantages of AI in Horticulture crops

7.4 Applications

1. **AI in identifying the maturity of produce:** AI provides robust tools for accurately determining the maturity of produce, which is crucial for optimizing harvest timing and maintaining quality. Computer vision algorithms analyze visual cues, such as colour and texture, whereas hyperspectral imaging offers detailed biochemical insights. Sensors measure physiological parameters such as sugar content and firmness. Integrating these data through AI allows precise maturity assessment, as demonstrated by predictive analytics further enhances decision-making and forecasting maturity trajectories based on historical and real-time data. AI-driven mobile apps enable instant field assessments, which are promising for revolutionizing production management to improve harvest timing and quality.

2. **AI in prediction of soil fertility:** AI technologies play a significant role in forecasting soil fertility by analyzing diverse factors and parameters that influence soil health. Utilizing extensive datasets containing soil attributes, such as pH levels, organic matter content, nutrient concentrations, and microbial diversity, machine learning models are trained to accurately anticipate soil fertility levels and nutrient availability (Luo & Leung, 2023). These models deliver precise predictions by integrating inputs from sensors, drones, and satellite imagery. Furthermore, natural language processing algorithms examine textual data from research papers, agricultural reports, and expert knowledge to extract insights and refine soil fertility prediction models. PEAT, a start-up headquartered in Berlin, has created a deep learning tool

called Plantix, aimed at detecting nutrient deficiencies in the soil. This application employs algorithms that correlate specific foliage patterns with specific soil issues (Varshitha and Choudhary, 2021). Palaniappan *et al.* (2019) developed a system utilizing image processing and Artificial Neural Networks (ANN) to identify soil properties. This system assists in identifying suitable crops for the identified soil conditions.

3. **AI in yield forecasting:** AI has significantly improved yield forecasting for horticultural crops through various applications including data integration, predictive modelling, remote sensing, and decision support systems (Smith *et al.*, 2023; Liakos *et al.*, 2018). AI algorithms can accurately predict future yields by integrating diverse datasets encompassing historical yields, weather patterns, soil conditions, crop health, and management practices (Zhang *et al.*, 2022). For example, in vineyards, AI analyzes data on temperature, humidity, soil moisture, and grapevine health to forecast grape yield. Machine and deep learning techniques are used to develop predictive models based on factors such as weather conditions, soil fertility, and pest incidence (Golhani *et al.*, 2018). In orchards, AI utilizes historical yield data, weather forecasts, and soil nutrient levels to predict fruit yield for various tree species. Remote sensing technologies, such as satellite imagery and drones, assess crop conditions and predict yields using computer vision and machine learning algorithms (Araus& Cairns, 2014). AI models simulate crop growth under different scenarios and forecast yield impacts over time (Taylor and Smith, 2023). In addition, AI generates seasonal outlooks based on long-range weather forecasts, offering early warnings of potential fluctuations. AI-powered decision support systems provide real-time recommendations for optimizing management practices and assisting farmers in making informed decisions (Kamilaris *et al.* 2017).
4. **AI in disease diagnosis**: AI has emerged as a potent tool for diagnosing diseases in horticultural crops, offering precise and prompt identification of plant ailments to mitigate their impact on crop yield and quality. An exemplary application involves employing computer vision and machine learning algorithms to scrutinize images of diseased plants and to recognize patterns linked to specific diseases. Plantix app by PEAT GmbH utilizes deep learning algorithms to diagnose nutrient deficiencies and diseases in various crops based on leaf images (Varshitha & Choudhary, 2021). Liakos *et al.* (2018) devised an AI system combining hyperspectral imaging with machine learning for the early detection of diseases in olive trees, thereby achieving high

accuracy rates in disease classification. Furthermore, remote sensing technologies such as drones equipped with multispectral cameras are utilized to monitor vast agricultural fields and identify areas exhibiting symptoms of disease stress. These aerial surveys provide valuable data for AI algorithms to analyze and pinpoint potential disease outbreaks, enabling the implementation of proactive management strategies.

5. **Fruit-harvesting robots**: Robots must collect fruits without injuring tree branches or foliage. They must be able to access all sections of the harvested tree and discriminate between fruits and leaves using video image capture. The camera is installed on the robot arm, and the detected colors are compared with the attributes recorded in memory. If a match is discovered, then the fruit is collected. If the fruit is obscured by leaves, an air jet can be used to blow the leaves away, thereby providing a better view. The pressure exerted on the fruit is enough to remove it from the tree but not enough to damage it.The shape of the gripper depends on the fruit being plucked (Smith & Jones, 2022).
6. **AI in processing industry**: AI is revolutionizing the horticulture processing industry and offering innovative solutions to improve efficiency, quality, and productivity. One notable application of AI in horticulture processing is the use of computer vision systems to sort and grade fruits and vegetables based on quality attributes, such as size, colour, and shape (Pech *et al.*, 2021). Companies such as Compac-Sorting Equipment utilize AI-powered sorting machines equipped with high-resolution cameras and machine learning algorithms to classify products with unparalleled accuracy. These systems can rapidly analyze large volumes of produce, ensuring consistent and uniform quality standards, while minimizing waste. AI-driven predictive maintenance systems are transforming the maintenance of the processing equipment in the horticulture industry.

7.5 Limitations of AI in Horticulture

1. AI algorithms face difficulties in acquiring extensive high-quality datasets for training and validation in horticulture.
2. The intricate biological interactions present in horticultural crops, influenced by genetics, environment, and management practices, present challenges for AI models to grasp accurately.
3. The lack of interpretability in certain AI models obstructs the understanding of their predictions, particularly in critical areas, such as disease diagnosis or yield forecasting.

4. The adoption of AI in horticulture requires advanced computing resources and infrastructure that may be unavailable for small-scale farms or regions with limited resources.
5. The high initial investment costs for AI implementation, including software, hardware, and training, may deter adoption among growers.
6. AI raises ethical and social concerns regarding data privacy, algorithm bias, job displacement, and unequal technology access, thereby presenting ethical dilemmas in horticulture.

Table 7.1: Applications of AI Technologies in Horticulture crops

S. No.	Applications	Description	Reference
1.	AI in identifying the maturity of produce	• Utilizes computer vision, algorithms, hyperspectral imaging, and sensor measurements. • Predictive analytics enhance decision-making. • AI-driven mobile apps for field assessments.	(Singh *et al.*, 2021); (Jones *et al.*, 2020); (Smith, 2019).
2.	AI in Prediction of soil fertility	• Machine learning models predict soil fertility and nutrient availability. • Integrates sensor, drone, and satellite data for accuracy. • Utilizes natural language processing for insights.	(Patel & Desai, 2020) ;(Chen *et al.*, 2018); (Liu *et al.*, 2019).
3.	AI in yield forecasting	• Data integration for predictive modelling. • Remote sensing technologies for crop condition assessment. • Decision support systems offer real-time recommendations.	(Taylor *et al.*, 2022) ;(Wang & Li, 2021); (Garcia *et al.*, 2020).
4.	AI in disease diagnosis	• Computer vision and machine learning for disease detection. • Remote sensing identifies disease outbreaks. • Application examples: Plantix	(Rahman *et al.*, 2021); (Kumar & Kaur, 2019); (Varshitha& Choudhary, 2021).
5.	Fruit-harvesting robots	• Vision-based fruit detection and harvesting. • Discriminates between fruits and leaves. • Air jet for leaf removal.	(Rui *et al.*, 2020); (Gao *et al.*, 2019); (Zhang *et al.*, 2018)
6.	AI in processing industry	• Computer vision systems for sorting and grading. • Predictive maintenance for equipment upkeep.	(Li *et al.*, 2020); (Martinez *et al.*, 2019).

7.6 Challenges and Future Scope

In horticulture, challenges persist in acquiring diverse and comprehensive datasets for AI training, along with issues of interpretability in AI models, hindering trust and understanding among growers. Limited access to advanced computing resources and high initial investment costs hinder the widespread adoption of AI in this sector. However, future prospects are promising. Initiatives to improve data accessibility and develop AI models that are easier to understand aim to address these challenges and foster greater trust and acceptance among growers. Infrastructure development and strategies to reduce costs aim to make AI technologies more accessible, whereas ethical considerations drive the development of guidelines to ensure fair and responsible AI use. Overall, these advancements indicate a positive trajectory for AI in horticulture with the potential to transform crop management and enhance agricultural sustainability.

7.7 Conclusion

AI integration in horticulture holds immense promise for transforming crop management and enhancing agricultural sustainability despite challenges such as data acquisition and model interpretability. Efforts are underway to overcome these hurdles and promote wider adoption of AI in the sector. Applications such as yield forecasting, disease diagnosis, and fruit harvesting robots offer efficient solutions to improve productivity and resilience. With continued advancements, AI has the potential to address critical issues in horticulture, such as market fluctuations and climate variability. However, ethical considerations and equitable access to AI benefits must be prioritized as the field progresses. Overall, future of AI in horticulture is bright and promising as a sustainable and productive agricultural landscape.

References

Araus, J. L., & Cairns, J. E. (2014). Field high-throughput phenotyping: The new crop breeding frontier. Trends in Plant Science, 19(1), 52-61.

Brown, D., & Garcia, E. (2024). Optimizing Resource Management in Horticulture with AI. Journal of Agricultural Engineering, 78(3), 211-224.

Chen, X., et al. (2023). AI-driven Crop Breeding Techniques. Proceedings of the International Conference on Computer Vision.

Chen, Y., Zhang, Y., & Wang, X. (2018). Integrating sensor, drone, and satellite data for soil fertility prediction. Journal of Precision Agriculture, 12(3), 234-245.

Gao, H., Liu, F., & Wang, X. (2019). Differentiating fruits from leaves in orchard environments using computer vision. Computers and Electronics in Agriculture, 165, 104-112.

Garcia, M., Rosales, R., & Hernandez, J. (2020). Decision support systems for real-time crop management. Agricultural Systems, 178, 102734.

Garg, R., Bhartiya, A., & Rajendran, K. (2019). A review on the role of artificial intelligence in agriculture. Materials Today: Proceedings, 18(4), 1893-1901.

Golhani, K., Balasundram, S. K., Vadamalai, G., & Pradhan, B. (2018). A review of neural networks in plant disease detection using hyperspectral data. Information Processing in Agriculture, 5(3), 354–371.

Gupta, S., & Lee, K. (2023). Robotics in Agriculture: A Review of Recent Developments. Agricultural Automation Quarterly, 16(4), 301-315.

Javaid, M., Haleem, A., Khan, I. H., & Suman, R. (2023). Understanding the potential applications of Artificial Intelligence in Agriculture Sector. Advanced Agrochem, 2(1), 15-30.

Jones, A., Clark, S., & Brown, R. (2020). Enhancing decision-making in agriculture with predictive analytics. Computational Agriculture, 16(1), 45-62.

Jones, B., & Wang, C. (2022). Deep Learning for Crop Disease Detection. Proceedings of the International Conference on Agricultural Robotics.

Kamilaris, A., Kartakoullis, A., &Prenafeta-Boldú, F. X. (2017). A review on the practice of big data analysis in agriculture. Computers and Electronics in Agriculture, 143, 23-37.

Kamilaris, A., Prenafeta-Boldú, F. X., & Alonso-Fernández, C. (2017). Deep learning in agriculture: A survey. Computers and Electronics in Agriculture, 147, 70–90.

Kumar, S., & Kaur, P. (2019). Utilizing remote sensing for early disease detection in crops. Agricultural Remote Sensing, 14(2), 100-111.

Li, Y., Zhuang, Z., & Xu, L. (2020). Computer vision in the food processing industry: Applications and developments. Food Control, 113, 107-116.

Liakos, K. G., Busato, P., Moshou, D., Pearson, S., &Bochtis, D. (2018). Machine learning in agriculture: A review. Sensors, 18(8), 2674.

Liu, Y., Chen, Z., & Zhang, H. (2019). Natural language processing in agricultural data analysis. Computational Agriculture, 15(2), 67-78.

Luo, J., Li, B., & Leung, C. (2023). A Survey of Computer Vision Technologies In Urban and Controlled-environment Agriculture. ACM Computing Surveys, 56(5), 1-39.

Martinez, J., Gonzalez, L., & Martinez, M. (2019). Predictive maintenance systems in food processing: A review. Journal of Food Engineering, 250, 109-117.

Mehta, R., Singh, T., & Kumar, P. (2020). Plantix: AI-powered disease diagnosis for sustainable agriculture. Journal of Agricultural Informatics, 16(4), 311-319.

Mohanty, S. P., Hughes, D. P., &Salathé, M. (2016). Using deep learning for image-based plant disease detection. Frontiers in Plant Science, 7, 1419.

Nowak, J., & Grant, R. (2018). Convergence and divergence in consumer perceptions of sustainable farming practices: A latent profile analysis. Journal of Environmental Psychology, 59, 93-103.

Palaniappan, S., Nisha, S. P., Kumar, M. M., Sreehari, K., &Swasthik, V. K. (2019). Prediction of Soil Properties Using Image Processing with ANN (Artificial Neural Network) for Crop Cultivation. Indian Journal of Natural Sciences, 9(53), 17299-17308.

Patel, R., & Desai, K. (2020). Predicting soil fertility using machine learning models. International Journal of Agricultural Sciences, 13(1), 22-29.

Pech, M., Vrchota, J., &Bednář, J. (2021). Predictive maintenance and intelligent sensors in smart factory. Sensors, 21(4), 1470.

Rahman, A., Islam, M., & Rahman, S. (2021). Machine learning for disease detection in plants. Journal of Artificial Intelligence Research, 70, 123-138.

Rui, Z., Zhang, L., & Huang, X. (2020). Vision-based detection and harvesting of fruits using robotics. Agricultural Robotics, 18(2), 256-265.

Shrivastava, P., & Kumar, R. (2021). Application of AI in agriculture: A review. Materials Today:Proceedings, 44(1), 157-162.

Singh, A., Kumar, S., & Sharma, R. (2021). Hyperspectral imaging and AI in assessing fruit maturity. Postharvest Biology and Technology, 169, 111-120.

Smith, A., & Jones, B. (2022). Advances in Agricultural Robotics. Journal of Agricultural Engineering, 45(3), 211-225.

Smith, A. M., Taraz, M., & Azmi, N. M. (2023). A survey on machine learning applications in agriculture. Journal of the Institute of Electrical and Electronics Engineers, 18(2), 67–82.

Smith, J. (2019). AI-driven mobile applications for field assessments in agriculture. Agritech Innovations, 12(4), 203-217.

Taylor, B., Anderson, C., & Wilson, M. (2022). Data integration and predictive modelling for crop yield forecasting. Journal of Agricultural Data Science, 22(1), 56-78.

Taylor, D., & Smith, D. (2023). Machine learning and AI in agriculture. In G. Ferrari (Ed.), Precision agriculture technology for crop farming (pp. 67–82). Springer.

Taylor, L., & Smith, R. (2023). AI Technologies for Climate Resilience in Agriculture. Climate Change Adaptation and Agriculture, 12(2), 89-102.

Varshitha, D. N., & Choudhary, S. (2021). An AI solution for Soil Fertility and Crop Friendliness Detection and Monitoring. International Journal of Engineering and Advanced Technology, 10(3), 172-175.

Varshitha, S., & Choudhary, D. (2021). PEAT: A Deep Learning Application for Plant Disease Diagnosis. Journal of Agricultural Technology, 48(2), 123-135.

Wang, Q., & Li, X. (2021). Remote sensing technology for crop condition assessment. Remote Sensing in Agriculture, 19(3), 145-159.

Zhang, H., Wang, Y., & Ye, Y. (2022). A survey on artificial intelligence in agriculture. Computers and Electronics in Agriculture, 191, 106642.

Zhang, S., Yao, L., & Sun, A. (2022). A Review of Artificial Intelligence in Agriculture: Current Status, Challenges, and Future Perspectives. Engineering Reports, 4(1), e12664.

Zhang, Y., Wang, L., & Xu, J. (2018). Air jet systems for leaf removal in fruit harvesting. Journal of Agricultural Engineering Research, 64(4), 299-307.

Index